I0006113

GLOBAL COMMUNICATION AND TRANSNATIONAL PUBLIC SPHERES

PALGRAVE MACMILLAN SERIES IN INTERNATIONAL POLITICAL COMMUNICATION

Series editor: Philip Seib, University of Southern California (USA)

From democratization to terrorism, economic development to conflict resolution, global political dynamics are affected by the increasing pervasiveness and influence of communication media. This series examines the participants and their tools, their strategies and their impact. It offers a mix of comparative and tightly focused analyses that bridge the various elements of communication and political science included in the field of international studies. Particular emphasis is placed on topics related to the rapidly changing communication environment that is being shaped by new technologies and new political realities. This is the evolving world of international political communication.

Editorial Board Members:

Hussein Amin, American University in Cairo (Egypt)
Robin Brown, University of Leeds (UK)
Eytan Gilboa, Bar-Ilan University (Israel)
Steven Livingston, George Washington University (USA)
Robin Mansell, London School of Economics and Political Science (UK)
Holli Semetko, Emory University (USA)
Ingrid Volkmer, University of Melbourne (Australia)

Books Appearing in this Series

Media and the Politics of Failure: Great Powers, Communication Strategies, and Military Defeats
By Laura Roselle

The CNN Effect in Action: How the News Media Pushed the West toward War in Kosovo
By Babak Bahador

Media Pressure on Foreign Policy: The Evolving Theoretical Framework
By Derek B. Miller

New Media and the New Middle East
Edited by Philip Seib

The African Press, Civic Cynicism, and Democracy
By Minabere Ibelema

Global Communication and Transnational Public Spheres
By Angela M. Crack

GLOBAL COMMUNICATION AND TRANSNATIONAL PUBLIC SPHERES

Angela M. Crack

GLOBAL COMMUNICATION AND TRANSNATIONAL PUBLIC SPHERES
Copyright © Angela M. Crack, 2008.

All rights reserved. No part of this book may be used or reproduced in any manner whatsoever without written permission except in the case of brief quotations embodied in critical articles or reviews.

First published in 2008 by
PALGRAVE MACMILLAN™
175 Fifth Avenue, New York, N.Y. 10010 and
Houndmills, Basingstoke, Hampshire, England RG21 6XS
Companies and representatives throughout the world.

PALGRAVE MACMILLAN is the global academic imprint of the Palgrave Macmillan division of St. Martin's Press, LLC and of Palgrave Macmillan Ltd. Macmillan® is a registered trademark in the United States, United Kingdom and other countries. Palgrave is a registered trademark in the European Union and other countries.

ISBN-13: 978–1–4039–7521–8
ISBN-10: 1–4039–7521–3

Library of Congress Cataloging-in-Publication Data

Crack, Angela M.
    Global communication and transnational public spheres / Angela M. Crack.
        p. cm.—(Palgrave series in international political communication)
    Includes bibliographical references and index.
    ISBN 1–4039–7521–3 (alk. paper)
        1. International cooperation. 2. Communication, International.
    3. Telecommunication—Social aspects. 4. Information technology—Social aspects. I. Title.

JZ1308.C73 2008
303.48'33—dc22   2007023401

A catalogue record for this book is available from the British Library.

Design by Newgen Imaging Systems (P) Ltd., Chennai, India.

First edition: February 2008

10 9 8 7 6 5 4 3 2 1

Transferred to Digital Printing in 2012

For my Family

CONTENTS

Acknowledgments

This book has evolved from my PhD thesis (Crack, 2004). I am immensely grateful to my postgraduate supervisor, Professor Tony McGrew. Without his invaluable guidance and unstinting support, neither my doctorate nor this book would have been possible. I would also like to thank my examiners, Dr. Tony Evans and Professor Barrie Axford, for their instructive feedback. I have particularly appreciated the interest that Barrie has maintained in my work following my viva.

The ideas developed in this book owe a profound intellectual debt to Jürgen Habermas, Nancy Fraser, and James Bohman. Their bodies of work have been a constant source of inspiration. My thought has also been informed by the constructive criticism of many colleagues over several years. I am grateful to Dr. David Owen who reviewed an earlier draft, and to all those who attended the Cumberlund Lodge conferences for their insightful comments. The late Alan Grant was kind enough to invite me to present at the Centre for Democracy Studies at Oxford Brookes University. I would like to extend my appreciation to Dr. Pete Woodcock and his colleagues at the University of Huddersfield, who invited me to present at their departmental seminar series. I benefited from the helpful feedback of participants at both events.

Thanks to my colleague Mike Mannin who helped with my workload as I was approaching my deadline. Nicola Ronan was a tremendous help in compiling the index. Sincere thanks to Professor Philip Seib for having faith in the book, and Toby Wahl and Kate Ankofski for their patience and editorial assistance. Maran and the team at Palgrave Macmillan won my admiration for their consummate professionalism. I would also like to acknowledge the comments of the anonymous reviewer. My deepest gratitude is reserved for my family, friends, and colleagues for their wonderful moral support, and not least for bearing my constant moaning about the "bloody book" with such good grace. There will now be a brief respite in whinging before normal service is resumed!

CHAPTER 1

Introducing Transnational Public Spheres to International Relations[1]

Globalization is paradoxical: it enhances the prospects for both domination and emancipation. On the one hand, neoliberalism is concentrating material resources in a transnational elite, exemplified by the growing disparity in global wealth distribution. A geographically disparate band of individuals and transnational corporations (TNCs) form an identifiable nucleus of significant structural power in current world order. Herman and Chomsky (1988) argue that this hegemony is partly sustained by the "manufacturing of consent" amongst citizens by a handful of global media conglomerates. Yet on the other hand, globalization has facilitated resistance by offering new opportunities for transnational dialogue, cross-cultural engagement, and grassroots political participation. It is often suggested that media and migration trends have weakened the role of the state in people's political horizons in favor of a growing cosmopolitanism (e.g., Held, 2003). This is partly evidenced by increasing participation in international civil society initiatives, such as the anticorporate movement, or the Make Poverty History campaign. Globalization, then, may be conceived as a dualistic phenomenon, a process of dynamic tension between oppositional tendencies.

A unifying theme in these trends is the pivotal role of information and communication technologies (ICT). The growth of neoliberalism has been largely fostered by the technological capacity to access information, transmit goods, and participate in distant financial transactions. ICT allow businesses to globalize their production, distribution, and marketing strategies. They enable capital to be transferred in immediate response to developments in the global marketplace, with minimal heed to political geography. ICT have become icons of modernity, generating the latest stage of capitalist development: the "knowledge economy." ICT are also implicated in the ideological

hegemony that sustains prevailing world order. The twin developments of digital convergence and deregulation have facilitated the
emergence of an oligopolistic global media, with unprecedented
reach, and with negative consequences for diversity of expression.
Simultaneously, ICT can be interpreted as potentially destabilizing
agents to this order. Applications such as the Internet enable transnational dialogue to flourish and countervailing political forces to
mobilize. Grassroots movements have harnessed the potential of the
Internet to forge solidarities, organize displays of resistance, and
articulate alternative visions of world order. Indeed, electronic communication forms a critical plank in the campaign strategy of transnational movements, with almost all having some type of online
presence. Of course, international political mobilization predates the
Internet, but the growing visibility of networked civil society demonstrates how ICT have precipitated an explosion in transnational communication. Therefore, the role of ICT in world politics is ambiguous:
it helps to bolster the prevailing order *and* allows contradictions in
the status quo to be exploited by counterhegemonic forces.

This dialectical tension may be usefully conceived with respect to
Habermas' distinction between "system" and "lifeworld." ICT can
serve systemic imperatives as well as give expression to transformative
potentialities in the lifeworld. So how can we evaluate the relative
influence of each opposing tendency? This interesting problematique
has not received due attention within International Relations (IR),
despite widespread consensus about the integral role of ICT in current global transformations. Notwithstanding important exceptions,
IR has largely neglected the analysis of epochal shifts in the way in
which we communicate.

This reflects the way in which the discipline tends to frame debate,
as indicated by the rubric of "International Relations." Traditionally,
those theories that focus on interstate war are favored, and alternative
approaches are marginalized. As Steve Smith argues: "it is very difficult to challenge that definition of the core problems of the discipline
without placing oneself *outside* the discipline. Thus, those approaches
that do not start with both *interstate* relations and with *war* are axiomatically placed in a defensive position with regards to their fit
within the discipline" (S. Smith, 2000: 378, original emphasis).
However, in an increasingly globalized and complex world, IR faces
the challenge of adopting a broader theoretical perspective in order to
retain topical relevance.

Accordingly, this inquiry presents a way in which global communication issues can be analyzed in an international relations[2] context.

It critically assesses the interaction between global media, sites of political authority and international civil society in order to evaluate the extent to which ICT can contribute to the democratization of world politics. It poses the following question: do the technological, political and social conditions exist for the emergence of transnational public spheres? Hence, the issues that are conventionally assumed to be in the province of IR, such as sovereignty and world order, are fused with a complementary focus on social movements and the political economy of the media.

The study of ICT has largely taken place outside of IR; likewise public sphere theory has been seen to have little application in an international context. This book challenges and explores the reasons behind these assumptions. It is useful to briefly review the history of IR in order to locate this inquiry in relation to the wider literature, and to illustrate the heuristic value of a public sphere approach.

1.1 THE FOUNDATIONS OF THE DISCIPLINE

IR is a young discipline. Its origins can be traced to the establishment of a chair in International Relations in 1919 at the University of Aberystwyth, Wales. The creation of this new subject area signaled formal recognition of a convergence of interests between the fields of Diplomacy, International Law, and History. The first generation of IR scholars staked out issues of interstate war and peace as a defining part of their disciplinary territory. Of course, these issues were far from alien to academia, but the lack of a dedicated research base or professional posts were thrown into sharp relief during the First World War. Aberystwyth's precedent was soon imitated by universities elsewhere in response to the atmosphere of heightened concern about international affairs. In the traumatized interwar years, many perceived no problem as more pressing than how to prevent the reoccurrence of a similarly horrific global conflict in the future. It is therefore not surprising that IR was initially dominated by idealist scholars (Zacher and Matthew, 1995).

Amongst the IR intelligentsia, it was commonly understood that the perennial threat of war was a product of "international anarchy." It seemed commonsensical that stability was elusive when there was no political authority at interstate level that could impose sanctions on violent aggressors in a way comparable to national governments. The League of Nations served as a repository for idealist hopes for a more peaceful future. Idealists argued that such a system of collective security accompanied by increased transparency in diplomatic practice

4 ◆ GLOBAL COMMUNICATION AND PUBLIC SPHERES

could mitigate the destabilizing implications of international anarchy (Doyle, 1986). However the events of the Second World War seemed to discredit idealist philosophy and encouraged a more pessimistic view to prevail. As a result, classical realist theory assumed a long unchallenged position of dominance in the discipline. Realists argued that interstate competition and hostility was an unavoidable consequence of the inherently insecure international environment (Donnelly, 2000). They posited that self-interested state actors are primarily concerned with the acquisition and maximization of power, and are distrustful of their rivals. Hence, war is endemic to international politics. Seminal texts such as those by Carr (1939) and Morgenthau (1948) established certain concepts as key to realist principles, such as the sovereign nation-state, the national interest, and the balance of power. These concepts effectively constituted the limits of inquiry in IR, to the extent that "In the first two decades after the Second World War the discipline and realism were widely regarded as one and the same thing" (Burchill, 1996: 80).

The term "realism" tacitly implies a shrewd objectivity, an impression that realists were keen to reinforce. They derided idealists as sentimentalists whose judgments were skewered by their normative beliefs. In contrast, realists portrayed their interpretation of world politics as pragmatic analyses grounded in concrete fact. In a highly influential article, Martin Wight appeared to support realist precepts by claiming that the international realm is characterized by recurrence and repetition, hence, is incompatible with progressive theory. For Wight, this explained why there was no equivalent body of knowledge in IR to that of Political Theory, as the former only relates to the "theory of survival," while the latter is concerned with the "theory of the good life" (Wight, 1966). However, Wight and his adherents failed to recognize that realism was as value-laden as its idealist counterpart. The epistemology of realism is constructed on subjective evaluations about which features of the international system are worthy of analysis (Rosenburg, 1990). In many ways, idealism and realism make the same value-judgments about the key actors and concerns of world politics.

The early focus on the causes and prevention of war influenced emerging understandings about "legitimate" subjects of study in IR. These preoccupations guided the way IR scholars framed their inquiries, the methods by which they conducted their research, and the conclusions they reached. Notwithstanding their differences, the two main schools of thought agreed on the broad definition of the intellectual domain of IR: the geopolitics of nation-states (Smith, 1995).

The discipline is no longer characterized by such consensus. Since Wight identified a paucity of international theory, an array of competing discourses has emerged to challenge the orthodoxy. A new, rich palette of actors, processes, and phenomena now supplement the staple fixations on the nation-state and war. Scholars such as Keohane and Nye (1977), from the pluralist school, and Gunder Frank (1980), from the neo-Marxist/dependency school, have led the way in this regard. The pluralist perspective presented numerous challenges to core realist assumptions. Pluralist international analysis was not biased toward nation-states; it recognized the political and economic significance of actors such as TNCs and intergovernmental organizations. Keohane and Nye (1977) described the interaction and interrelations between these actors in terms of an international web of complex interdependence. Pluralists also disputed the supremacy that realism accorded military force in international affairs, and instead depicted power resources as multidimensional. They questioned the realist focus on international conflict, given the higher incidence of multilateral cooperation. Instead, it was surmised that the international realm was better understood as an "ordered anarchy" (Zacher and Matthew, 1995). Regimes became a focal issue of pluralist research, defined as "sets of implicit or explicit principles, norms, rules, and decision-making procedures around which actors expectations converge in a given area of international relations" (Krasner, 1982: 186). Many pluralists were intrigued at the possibilities that regimes represented for greater international integration and policy coordination.

This revived normative interest (though muted) in the potential for progressive international politics created an intellectual climate that was receptive to the introduction of Marxist-influenced theory. Dependency theorists such as Andre Gunder Frank (1980) argued that the relationship between the Northern capitalist states and the Southern peripheral states was inherently exploitative. According to this perspective, poverty in the global South is not due to "underdevelopment" as often claimed by the Northern elite. Rather, the development discourse is an ideological smokescreen that disguises the real systemic cause of inequality: the structural conditions of transnational capitalism. It was argued that the fate of Southern countries was effectively dependent on a global economy that serves the interests of the rich. Further, world-system theory suggested that international politics could be conceptualized in terms of class, and analyzed the way in which the global capitalist system institutionalizes inequalities between the privileged "core" and the marginalized "periphery" (Wallerstein, 1977).

The development of pluralism and neo-Marxism subjected realism to its most significant challenge since its inception. Kenneth Waltz reformulated the main tenets of realism and ensured the continuing relevance of the discourse in his seminal work, *Theory of International Politics* (1979). Waltz's neorealist approach assumed an intellectual hegemony over the discipline that still retains considerable influence today. He reasserted the position of nation-states as key players, and propounded the importance of understanding the operation of power in the international system. For Waltz, recurrent conflict was structurally determined by international anarchy. He reasoned that this insecure environment instills nation-states with an instinct for self-preservation and an intrinsic mistrust of their rivals. Therefore, these systemic imperatives produce similarities in the foreign policy behavior of states, regardless of differences in internal composition. Waltz and his adherents represented a decisive return to the discipline's narrow and militaristic focus, leaving little room for dedicated analyses on the social and political import of communication.

A notable exception is Karl Deutsch, who spearheaded an early attempt to reconcile theories of communication with IR as far back as the late 1950s (e.g., Deutsch, 1957, 1963, 1966). Deutsch was interested in communication flows as indicators of levels of social integration, and investigated the possible emergence of a European security community. Although there are aspects of interpretivist analysis in his work, Deutsch was generally biased toward positivism, arguing that it was only through quantitative data that facts could be ascertained and evidence divined. Thus, he prioritized communication flow as a measurable variable, thereby demonstrating how different national communities can be identified through concentrated clusters of communication patterns (such as the density of postal or telephone exchange). He suggested that the irregularity of this distribution partially explains the development of nationalist sentiment, and hypothesized that community could be expanded with broadening of communication flows. His influence can be discerned in the subsequent work of other IR theorists such as Hedley Bull (1966).

Deutsch's work was pioneering not just because of the topic but also because it contained a distinctive normative element, which was unusual in the context of a discipline in thrall to realism. But Deutsch's faith in positivism led him to make a simplistic equation between increased communication and social integration. It is surely naïve in the extreme to assume that the expansion of communication will necessarily result in closer social bonds and the forging of common identities. For example, much international communication is merely instrumental, and only requires minimal cross-cultural understanding—such as routine business transactions. In other cases, exposure to a different culture may stoke rather than quell nationalist or fundamentalist sentiment. A qualitative analysis of communication flows provides an incomplete picture of the actual *quality* of interaction. What is also needed is a complementary focus on message content. However, post-Deutsch, IR scholars continued to neglect communication issues for years, and so these issues remained unsatisfactorily unresolved. It is interesting to consider why this was the case.

It is evident that since the discipline was established, the majority of IR scholars have not perceived communication issues as germane to world politics. The first "great debate" in IR between idealism and realism was underpinned by an ontological and epistemological consensus, which was further reinforced by the emergence of neorealism (Knudson, 1992). Nation-states were understood to be the key actors of the international system and differences in their internal composition were considered largely irrelevant. States were portrayed as self-contained, unitary actors, and war and state interaction as the most important characteristics of the global political process. Intellectual inquiry in IR has largely been characterized by the search for ahistorical "truths" in the international system, rather than the search for the transformative. As Ruggie observes, IR theorists are not "very good…at studying the possibility of fundamental discontinuity in the international system" (Ruggie, 1993: 143–144). Those that do attempt to account for change usually focus on the mode of production or military capability as the most important variables (e.g., Wallerstein, 1977; Carr, 1939). If considered at all, communication technologies are usually subsumed under or interpreted with reference to these key factors. Thus, the conventional preoccupations of the discipline have precluded examination of the different ways that communication developments impact world politics.

Since then, postpositivist theories have gained increasing prominence in IR, albeit belatedly compared to other branches of social science. Postpositivism submitted a comprehensive critical assessment of conventional perspectives on theoretical, epistemological, methodological and normative grounds. It exposed IR orthodoxy as biased toward system-maintenance. This conservative standpoint represents a value-bound understanding of the world that belies the positivists' oft-proclaimed commitment to objectivity (George, 1994: 11; Cox, 1981: 130). Neorealists may protest loudly about the supposed neutrality of "scientific" behavioralist analysis, but actually the neorealist worldview effectively reifies the status quo by accepting the prevailing order as a given. Even seemingly critical perspectives such as dependency theory are often underpinned by a similarly conservative outlook. Critique alone is inadequate if one does not want to be complicit in the reification of global inequality; it is incumbent upon the theorist to identify contradictory tendencies that could lead to future social change (Linklater, 1990a: 148). Otherwise, theory has a stabilizing effect on extant social order, whether this is intended or unintended. Those who are most privileged by this order are the greatest beneficiaries of this stabilizing function (Cox, 1981: 129; Peterson, 1992: 14–15). Such claims were made in distinct ways by three main schools of thought that shared a common radical agenda: feminism, critical theory, and postmodernism.

1.2   POSTPOSITIVIST THEORIES OF IR

Two disclaimers are immediately required before proceeding. First, the use of generic names to classify highly individualistic theoretical approaches is problematic. There is usually significant divergence of opinion in all schools of thought, not least in branches of postpositivism. This is especially so with regard to "postmodernism," where some of the theorists that are usually subsumed under that label explicitly defy the notion of categorization. Nevertheless, generic labels persist for much the same reason as they are employed here: the sake of convenience. Second, the review that follows also serves as a reflective exploration of my own theoretical sympathies. Feminism and postmodernism have made vital contributions to IR, but I contend that it is not clear that either can address the concerns that are assimilated by a sophisticated critical theory. International critical theory develops a normatively motivated account of existing social conditions and explores the potentialities for emancipatory transformation. Postmodernism encompasses a similar approach, but is unsatisfactorily vague on questions of how to critically engage and transform the existing world order. Feminism offers an important corrective to masculinist standpoints, but can also neglect analysis and critique of exclusions based on categories other than gender. However, the insights of both postmodernism and feminism can contribute to the development of a reflexive critical theory, as shall become clear.

Feminist theory began to make a notable impact on the discipline in the 1980s, emerging as a response to the exclusion of women in conventional IR discourse. As Jean Elshtain has remarked, in IR theory, "what gets left out is often as important as what is put in and assumed" (Elshtain, 1995: 41). Conventional theory is not adequately framed for the purpose of understanding the social and historical constructions of gender, the modern state and state system; nor does it recognize the agency of people as "women" (Tickner, 1992a, 1992b). Instead, "malestream" discourses are methodologically biased toward individualism and inductive reasoning. Hence, gendered hierarchies are taken for granted; and gendered dichotomies are presumed to be unproblematic. For feminists, theory that fails to recognize the significance of gender as a tool of analysis performs the political act of reproducing the status quo (Peterson, 1992; Enloe, 1990; Zalewski, 1994). It is argued that deconstructing the exclusionary ontological and epistemological foundations of IR will enhance our understanding of world politics. Critics point out that inequalities are multifaceted, and focus on gender may distract attention from other disparities and abuses of power.

Critical theory was introduced to IR by Robert Cox's pathbreaking 1981 essay, "Social Forces, States and World Order." Cox described contending theories as falling into two categories: "problem-solving theory" and "critical theory." "Problem-solving theory" referred to the typical positivist approach, which "takes the world as it finds it, with the prevailing social and power relationships into which they are organized, as the given framework for action" (Cox, 1981: 128). Problem-solving theory concentrates its efforts into finding ways to eliminate certain types of threats to the continued operation of world order; it does not subject this order to robust challenge. Thus, it effectively reifies and legitimizes the status quo (ibid.: 128–129). Cox contrasted critical theory in several ways. Critical theory recognizes that all conceptual frameworks can be located in a particular time and place. In other words, social theory is essentially subjective and shaped by social, cultural, and ideological influences. The critical theorist will attempt to bring these concealed perspectives to consciousness and reflect upon how they impact upon the task of philosophical inquiry. Critical theory exposes the fallacy of "objectivity" in social analysis, and takes an explicitly normative stance in favor of human emancipation. It is oriented by an interest in exploiting the inherent potentialities in the present for societal transformation. Cox's writings are heavily influenced by Antonio Gramsci, particularly with regard to the concepts of production, the state, social forces, and hegemony (Cox, 1983, 1987). Cox reconstructed Gramscian hypotheses to theorize the rise and decline of a "historical structure." This is defined as a certain configuration of three categories of forces: ideas, material capabilities, and institutions (Cox, 1981: 141). Between each category there are reciprocal flows of influence. Cox argues that this model can aid the conceptualization of structural transitions in the past, which in turn will better enable one to assess the possibility of world order transformation in the future.

Postmodernism defies summary or definition by its very nature. It could be said that it is easier to define what it is *not*, rather than what it is. It is popularly described in negative relation to modernity, as an attitude of "incredulity toward metanarratives" (Lyotard, 1984: xxiv). Postmodern IR theory shares similar interests with international critical theory, such as opening up discursive spaces for self-reflection, and advancing the understanding of marginalized themes in world politics (e.g., Ashley and Walker, 1990; Der Derian and Shapiro, 1988; Walker, 1993). However, the points of divergence between both camps on questions of rationality, universalism, and ethics can be profound. For example, Rengger ironically observes that critical theory is as dependent on foundationalism and universalism as are "problem-solving" perspectives. He argues that critical theory forms part of a wider modernist discourse, as it holds that "the international order constitutes a dialectical process, but one presumably, like Marx's, with a fixed terminus: a telos to aim for and to bring about" (Rengger, 1988: 83). According to postmodern interpretation, rationality has not and will not be a discourse immune from power and domination—rather, rationality embodies these inequalities. In this sense, "to theoretically privilege one side of modern rationality…is to engage in the practice of exclusion (and sometimes terror) that is the experience of the other side—that which has no (rational) voice" (George, 1994: 161). Critical theory can be seen as complicit in the reproduction of unequal power relations.

Rengger's critique prompted a well-known response from Mark Hoffman, who argued that it was an ill-informed misrepresentation of a sophisticated paradigm. Hoffman argued that critical theory should not be dismissed so readily since it is "the most self-reflective outpost of the radical tradition of the Enlightenment," and used Habermasian analysis to reject Rengger's claims of underlying instrumentalism in critical thought (Hoffman, 1988: 92). For Hoffman, critical theory seeks to critique the universalization of a single form of rationality, "namely instrumental, economic and administrative reason," and so represents an open-ended and evolutionary rationality. Critical theory

> …retain[s] a concept of reason which asserts itself simultaneously against both instrumentalism and existentialism, which is exercised in conjunction with normative concerns and which is critically applicable

to itself. The essence of rationality, in the context of critical theory, entails a limitless invitation to criticism. In consequence a complacent faith in *rationalism* is ruled out. (ibid)

To paraphrase Hoffman's argument, the emphasis in critical theory on reflection and indeterminate knowledge act as safeguards from the worst excesses of Enlightenment determinism and foundationalism (Linklater, 1996a). Critical theory does not purport to ascertain "objective truths." This is the goal of instrumental-type rationality, which is oriented by an interest in control over nature and other human beings. Rather, critical theory understands the possible fallibility of *all* knowledge claims. It resists the teleological "fixed terminus" approach, the imposition of a futuristic utopian model of society upon others. Critical theory analyzes the counterdiscourse in modernity and recognizes that the limitations of instrumentalism can only be rectified by a reflective and open-ended rationality (Hoffman, 1993: 199).

Similarly, Hoffman argued that the sensitivity and reflexivity of a critical approach effectively assuages the dangers inherent to universalism. Critical theory both "recognizes the problem [of universality] and acknowledges its own limitations" (Hoffman, 1988: 93). The universality of critical theory is therefore "cautious and contingent," and respectful of cultural difference (ibid.). Nevertheless, the adoption of certain universalistic precepts is necessary for a social and political theory that seeks critical engagement with the world. As Hoffman observes: "The difficulty with the anti-universalism of radical interpretivism is that it offers us no reason to move in one social direction or another. We become dispassionate observers rather than concerned critics" (ibid.). Hence, "the radical interpretivist approach contains within it an element of conservatism and stops short of the aims of critical theory to change the way we talk *and act* in the world" (ibid.: 94). Critical theorists are explicitly motivated by an emancipatory interest in transformation, which enables them to be normatively engaged with social and political issues in a less ambiguous way than is possible for postmodernists.

Rengger's critique is useful in focusing attention on some possible pitfalls of critical theory—such as the dangers of progressivist, totalizing tendencies in Enlightenment discursive traditions. If critical theory is to be truly reflexive it should seriously acknowledge and engage with these concerns, or it will indeed be subject to the same shortcomings as problem-solving theory. However, the unique strengths of international critical theory are elegantly defended by Hoffman. Most importantly, he underlines how essential the contributions of feminist and postmodernist thought are to the development of an evolutionary rationality that informs the critical approach (Linklater, 1992: 85). Each discourse is linked by a common interest in challenging the reproduction of inequalities in world order. The fascinating debates that emerge between the schools of thought usually enrich understanding on both sides. Critical theory is best able to synthesize insights from complementary approaches and so is well designed to comprehensively analyze the multifaceted complexities of global power relations. Not least, these strengths are derived from critical theory's unequivocal commitment to the "emancipation of the species" (Linklater, 1990a: 8). Critical theory should not be abstract or disengaged from real events, but instead represent a political call for fundamental change.

The theories outlined above have had the most significant impact on the historical development of the discipline, but this is by no means an exhaustive account. Since the 1980s, the customary focus and traditional boundaries of IR inquiry have been subjected to sustained challenge. IR is now a considerably eclectic field of study. In recent years, numerous rival theoretical perspectives have garnered popularity: from green theory (e.g., Hurrell and Kingsbury, 1992; Chatterjee and Finger, 1994) to social constructivism (e.g., Checkel, 1998; Wendt, 1999).

Postpositivist approaches have introduced contemporary social science debates to IR, and ensured that previously marginalized questions of ontology, epistemology and methodology are now central concerns of the research agenda. The alternative theorists also disputed the neorealist assertion that international relations operates by a discrete set of rules, and so requires discipline-specific forms of study. Critical theorists have instead explored the application of social theory and historical sociology for international analysis (Booth, 1995: 119). Feminists demonstrate the effects of patriarchic structures from the intimate to the international level (e.g., Sylvester, 1994; Enloe, 1990). Postmodernists protest that disciplinary boundaries are a tool of intellectual oppression (e.g., Der Derian and Shapiro, 1988; Ashley and Walker, 1990). The study of what is still anachronistically called "International Relations" has been transformed into a subject infinitely more broad and diverse by these radical reinterpretations. Perhaps "Global Politics" would now be a more appropriate title (Cox, 1992: 132). After a long period of intellectual hegemony, the widespread subversion of governing orthodoxies is perhaps indicative of a common feeling of disorientation caused by the turmoil of globalization (Rosenau, 1990). But it is surprising that few significant studies on ICT have emerged from the postpositivist turn, despite the centrality of communicative themes in each of these perspectives.

In other branches of social science, postmodernists and feminists in have written extensively about ICT and global media. Postmodernists have been intently interested in the new prospects that communication developments open up, such as opportunities to realize a "virtual reality," to transcend the restrictions of time, space, and the body, to subvert one's identity; as well as more sinister possibilities for increased government surveillance (e.g., Burnett, 1995; Baudrillard, 1995). Feminists have examined the unequal access and ownership of communication resources between the sexes, and how media portrayals of women reinforce gendered power relations (e.g., van Zoonen, 1991). Unfortunately, the insights of both approaches do not tend to be fully integrated with an explicit IR frame.

Perhaps the most obvious omission is that of the critical theorists. For example, Gramsci's concept of hegemony is central to the work of the "Italian School," involving issues of information and the production and development of social knowledge. There are few instances of these themes being developed satisfactorily, particularly with regard to ICT. Cox provides the most sustained effort in his "historical structures" approach (Cox, 1981: 136). He conceptualizes "ideas" as part of a dialectical triad of "categories of force" within a historical structure. "Ideas" refer to both intersubjective meanings and differing perspectives, and form a key component in Cox's conceptualization of the forms of domination that govern the reproduction and eventual decline of a given world order. However, there is a general undertheorization of process in Cox's work, and this is partly because he has no specific theory of communication, and as a result he delivers an incomplete account of the transition between one historical structure to another. Cox's conceptualization of "ideas" does not stretch as far as to consider issues of the storage, transmission and distribution of knowledge, and how fundamental transformations in each of these categories will affect the range of possible alternatives (Comor, 1994). The result is ambiguity about how change will occur.

Recently, there have been encouraging signs that this curious lacuna in the IR literature is beginning to be addressed. Over the past few years, a number of texts have accredited ICT with a significant role in world politics. For instance, there has been increasingly frequent mention of ICT as an important corollary of globalization (e.g., Clark, 1997; Scholte, 2000; Aronson, 2005). There have even been several undergraduate-style textbooks published in the field that focus exclusively on ICT. Although worthwhile, these studies tend to provide more of an overview to the subject rather than establish a coherent theoretical position (e.g., Chadwick, 2006; Frederick, 1993; Mohammadi, 1997; Thussu, 2006). Likewise, there are authors who write on international law, and adopt a policy-orientated perspective—such as Cees J. Hamelink (1994, 1998), who examines prevailing political practice and international law in areas such as transborder data flow, international broadcasting, and the standardization of consumer electronics. Amongst IR theorists, the influence of ICT has been interpreted in varying ways: either to support claims of the decline of the state owing to growing global economic and political interdependence, or to argue that state power has been enhanced by new surveillance and data-processing capabilities. Examples include James Rosenau (1990), for whom ICT are key to explaining growing turbulence in world politics, and Susan Strange (1996), who uses changes in communication technologies to partly account for the rise of a networked global economy and the "retreat of the state." However, communication generally remains undertheorized in IR. The studies cited above provide a good example of how ICT tends to be treated in the literature—as an exogenous force, despite common agreement in other disciplines such as Communication Studies and Sociology that technology is inherently social and political (e.g., Thompson, 1995). The academic convention of disciplinary boundaries dictates that one must turn to the Media/Communication Studies literature for dedicated analyses of ICT. However, as self-contained subjects drawing from a specialized scholarship, these inquiries can be frustratingly limited. Typically, they fail to extend into issues that are of interest to the IR scholar, such as state sovereignty and world order. Hence, there is a potential nexus between IR and Media/Communication Studies that remains underdeveloped.

Postpositivist approaches are best placed for studies of ICT in world politics. International critical theory has an underdeveloped capacity for such research, and both postmodernism and feminism have the benefits of a preexisting, well-established body of work in other disciplines. Although critical IR theorists have been notoriously slow in incorporating ICT analysis in their research, there has recently been some interest in tapping this potential (e.g., Comor, 1994; Baynes, 2001). Indeed, it is largely within the critical-theoretical tradition that the most intriguing literature is to be found. In a notable series of publications, Linklater (1990a, 1990b, 1990c, 1996a, 1998) develops Habermasian theory as a philosophical alternative to positivist approaches, with dialogic communities providing the basis for world order. Scholars from complementary fields have also found public sphere theory a useful basis from which to explore the import of cross-border dialogue, resulting in a burgeoning literature (e.g., Bohman, 1998; Calhoun, 2003; Crossley and Roberts, 2004; Dahlberg, 2001; Sassi, 2001). Most recently, Lynch (1999, 2000), Mitzen (2001, 2005), and Brunkhorst (2002) have introduced public sphere terminology to IR. This book is an attempt to build on these contributions by developing the notion of a classic Habermasian public sphere. A refined Habermasian approach could offer a productive way to develop analyses of ICT in world politics that avoids the deficiencies and limitations of some of the literature discussed so far.

1.3   An Outline of the Main Thesis

The "public sphere" offers a useful theoretical framework with which to analyze and critique the relationships between forms of media, sites of political authority, and civil society actors. The concept describes a mode of discursive engagement about matters of governance, which secures the equitable involvement of the greatest number. In a public sphere, each participant can air his/her concerns and they are prepared to listen and engage with the concerns of others. The outcome of debate will reflect general concord, rather than the power differentials between participants. Likewise, the rational merit of argument will supersede the social and economic status of interlocutors. The value of a public sphere approach is that it enables precise and effective critique of situations that contribute to the reproduction of inequalities in world order. The public sphere is a normative ideal against which actual deliberative practice can be compared, deficiencies identified, and remedial action suggested.

It is important to clarify the distinction between the public sphere and civil society. They are closely related concepts—indeed, the intersection between the two is crucial—but the terms are not interchangeable. "Civil society" refers to non-state actors and organizations (Anheier et al., 2001: 4), whereas "a public sphere" describes a site of free and open discussion between civil society actors (Habermas, 1999: 23). Public sphere theory provides an appropriate theoretical framework for the purposes of this inquiry for two main reasons. First, it is centrally concerned with the relationship between politics and methods of communication, and therefore ideally placed for an exploration of the nexus between IR and Media/Communication Studies. Second, public sphere theory investigates the most appropriate form of social organization in which to develop effective political communication. It hinges on issues of participation and inclusion, and so enables the identification and critique of antidemocratic processes and repressive elements of civil society. It is therefore well designed for a critical-theoretical investigation into the role of ICT in contemporary global transformations.

Critics often charge that the emphasis on inclusion and rationalism in public sphere theory represents the worst aspects of Western universalism. In the interests of reflexivity, it is important to seriously engage with these concerns, and they have been foreshadowed in the Rengger/Hoffman debate outlined above. I discuss them further in section 2.3. It will suffice here to assert that the public sphere both recommends minimalist universalism and the safeguarding of diversity. A tension exists between these two orientations, but such tension may be inevitable in democratic normative theory. A public sphere approach can address these issues with sensitivity because it relates to the procedures and conditions of deliberation, not the substantive outcome of debate. Its ambitions are limited to setting guidelines for ensuring the widest possible inclusion in public life. Hence, it can balance a "cautious and contingent" universality with respect for difference (Hoffman, 1988: 93). In the remainder of this section, I want to outline the main steps in the argument put forth herein regarding the reconfiguration of public sphere theory.

The public sphere has been conceptualized in a variety of ways, including Arendt's theory of the agnostic public (1958), and Dewey's examination of American small town publics (1927). However, Habermas' classic formulation of the bourgeois public sphere is the best known and the most influential. With reference to the emergence of the eighteenth-century English public, Habermas gave a historically and sociologically grounded account of the possibility for human emancipation in communication in *The Structural Transformation of the Public Sphere* (1999, originally published 1962). Habermas conceived of the public sphere in a local/national context—so did Arendt and Dewey. Indeed, many other distinguished public sphere theorists also frame their versions of public spheres in bounded territories (e.g., Koselleck, 1988; Negt and Kluge, 1993; Ryan, 1992; Fraser, 1992). However, the huge growth in cross-border communication through ICT raises questions about the possible expansion of public spheres beyond the nation-state. The temporal-spatial barriers that were once constraints on distanced communication have become less relevant in today's media-saturated environment. Potentially, the material infrastructure exists to support a transnational deliberative public who can communicate in real time. Consequently, academic interest in the concept of transnational public spheres has been piqued. The rationale is beguiling. The agents and communicative structure of the state-bound public sphere (i.e., national citizenry, print media) seem to have been supplemented with transnational counterparts (i.e., international civil society, ICT).

However, as Fraser warns, we must be cautious of a lazy analogy between online political activity and the type of civic participation that characterized the bourgeois public sphere (2005). Consider the substantive differences between the domestic and international realms. First, the national mass media focuses on domestic affairs of state, which provides some coherence to civic deliberation. Compare this with a decentralized technology such as the Internet, which hosts fragmented discourse on a countless variety of forums. Second, state-based public sphere theory has developed on the basis of a direct relationship between a sovereign political authority and the public opinion of the citizenry. All citizens were bounded by a delimited territory and shared national identity. There are no such common denominators at a transnational level, neither are there clearly defined relations of political accountability. The state apparatus is essential to the national public—it provides theoretical coherence, political significance, social viability, and normative value. There is a need for further investigation as to whether the concept of the public sphere will bear translation to the transnational domain. This is not to suggest that the notion of a transnational public sphere has no analytical value, rather that there is an evident need for conceptual clarification (ibid.).

Few theorists systematically investigate the conditions of possibility for the emergence of transnational public spheres. Many assume that such a sphere or spheres already exist. But the traditional association of the virtual space of the public sphere with the physical space of the territorial nation-state cannot be so readily dismissed. Recasting the public sphere in an internationally anarchic environment has complex theoretical implications. Hitherto, these issues have received little consideration in the literature, resulting in ambiguity about the structural foundations of cross-border publicity. This inquiry sets out a methodical framework for the theorization of the institutional requisites of transnational public spheres. Further, it is oriented by an explicit interest in emancipation, and aims to redress the general neglect in recent public sphere literature toward normative issues. As Fraser contends, public sphere theory "is currently in danger of being depoliticized" (ibid.). The public sphere was conceived as a forum for grassroots political mobilization, yet some contemporary scholars have overlooked the deliberative activity of the citizenry in favor of governing institutions. For example, Lynch and Mitzen have reinvented the public sphere in terms of state actors, thus synthesizing the concept with IR concerns (e.g., Lynch, 1999; Mitzen, 2001, 2005). Although their work has value, they are in danger of diluting the radicalism of public sphere theory by failing to focus on how the populace can bring political authorities to account. To employ public sphere theory in a purely descriptive way is to deny the revolutionary purpose for which it was designed: as a means to critique the powerful and to promote democracy. The alternative approach that I propose is based on a reformulated version of Habermasian theory. The remainder of the book is structured as follows.

In the second chapter, Habermas' well-known account of the bourgeois public is used as an entry point into theorizing about current transformations of public spheres. *Structural Transformation* is an impressive blend of empirical research and critical-theoretical analysis, and is the mainspring for this inquiry. It must be emphasized that Habermas' claims regarding communicative rationality do not concern me here. Habermas' early public sphere theory was historicist— later it was supplanted by a more abstracted version influenced by discourse ethics (Habermas, 1984, 1987). The account herein is inspired by the situatedness of the former rather than the transcendentalism of the latter. However, Habermasian public sphere theory contains a number of widely acknowledged flaws, involving issues such as historical inaccuracies and the controversial distinction between public and private interests. Habermas himself has recognized the validity of many of these criticisms (e.g., Habermas, 1992a). But the account of the bourgeois public also rests on a number of statist presuppositions that are also evident in other reformulations of Habermasian theory (e.g., Fraser, 1992; Ryan, 1992; Meehan, 1995). These need to be reassessed because globalization raises a variety of questions about the contemporary status and structure of public spheres. For example, are ICT providing adequate discursive spaces for transnational democratic dialogue? Is the primacy of the nation-state as a site of sovereign power in decline? And can we identify the emergence of transborder citizen networks that engage in rational-critical deliberation?

In the third chapter, I posit a functional definition of a public sphere that acts as a theoretical probe into these questions. It is inspired by Habermas and based on a critical appraisal of recent public sphere theory. The definition is as follows: *a transnational public sphere is a site of deliberation in which non-state actors reach understandings about issues of common concern according to the norms of publicity*. The "norms of publicity" have several requirements. First, that debate should be free and open to all affected actors as nominal equals, regardless of their social status. Second, debate should be conducted according to certain principles. For example, participants should endeavor to make their contributions intelligible to others; and when interrogated, be willing to provide reasoned justification for their opinions. Third, arguments should be oriented toward understanding and adjudicated through rational judgment. This definition is designed to be flexible enough to accommodate the diversity that is typical when evaluating non-state based, cross-cultural communication. It also permits the existence of multiple spheres, as the transnational environment is host to a bewildering array of voices, groups, and interests. Note also that this definition neither presumes the locus of spheres nor does it pre-determine which issues merit public debate.

Three contemporary trends may provide the structural precondi-tions for emergent transnational public spheres. These are *transborder communicative capacity, transformations in sites of political authority,* and *transnational networks of mutual affinity.* They are modified versions of the prerequisites of the Habermasian public sphere. The conjunction of these structural preconditions would produce a suit-able environment for the emergence of transnational public spheres. Important qualitative requirements also need to be met, which are only possible if there are supportive institutions in each category. As in the Habermasian ideal, media should be free and open; governance structures should be accountable and receptive to public opinion, and civil society institutions should observe basic deliberative norms. If there is a convergence of these enabling conditions around a given issue-area, then transnational networks could host meaningful criti-cal dialogue. The subsequent chapters analyze whether each precon-dition can be identified in current world order.

The fourth chapter examines *transborder communicative capacity,* entailing all kinds of ICT. However, the main focus here is on new media, particularly the potential of the Internet. These networked technologies represent a qualitative difference from older forms of mass media in terms of the scope, structure, and speed of communication. ICT provide citizens with increased access to information and enhance opportunities for political mobilization across borders. However, wider participation and freedom of speech is threatened by the continuing encroachment of state and corporate power and by huge global dis-parities in access and ownership of media technologies.

The fifth chapter discusses *transformations in sites of political authority,* or the growing challenges posed to state sovereignty by evolving structures of global governance. Global governance describes the gamut of rules, regimes, and norms that constitute the control mechanisms for the management of transnational issues. Global governance is multilayered, spanning the local, regional, and the supranational. It encompasses a multitude of actors as well as states, including local authorities, international organizations, TNCs, nongovernmental organizations (NGOs), courts, and public and private regulatory bodies. The intense activity and expanding remit of these actors invites a reassessment of conventional assumptions about the primacy of the nation-state in world politics. Therefore statist presuppositions about public sphere deliberation are also called into question by possible changes in the global architecture of political authority.

The sixth chapter considers *transnational networks of mutual affinity*, which refers to civil society groups using ICT to communicate and engage in political activism across state borders. A public sphere depends on feeling of affinity amongst the interlocutors; otherwise people would attempt to further their interests by means other than rational argumentation. Following Dahlgren, the word "affinity" is used in a minimalist sense, meaning only that citizens recognize the moral-political validity of inclusive discourse (Dahlgren, 2002: 17). In other words, public sphere participants should demonstrate an implicit belief in the importance of interacting and exchanging views with one another, and a normative conviction that authority can only be legitimated through public opinion. These values are evidenced when "norms of publicity" are embodied in discourse (e.g., inclusivity, intelligibility, accountability, reflexivity). There are a number of factors that inhibit the emergence of critical publicity amongst transnational networks, such as geographic diffusion, lack of common citizenship, and the anonymous methods of communication that typify Internet discourse (Calhoun, 2003). Despite these complications, there is a seemingly unstoppable upsurge in transnational activism—and such popularity contrasts sharply with the faltering fortunes of mainstream political parties in established democracies.

A number of case studies provide microcosmic examples of how embryonic public spheres can arise through transnational coalitions of networked citizenry. These are drawn from three main subject areas: the international women's movement, the Zapatistas, and Greenpeace. Each subject area has been chosen because it exemplifies a different context from which the participants derive a sense of mutual affinity. In the case of the women's movement, the basis is gendered experience, in the case of the Zapatistas, it is anti-neoliberal rhetoric, and in the case of Greenpeace and other associated movements, it is the ecosystem. Effectual public spheres are not just sites of critique but are also sources of societal transformation. The import of a public sphere can be measured by the effect that dialogue has on the political agenda and the actual exercise of authority. The case studies will illuminate instances where activists have influenced hegemonic discourses, or made a perceptible difference to the international institutional framework. These points of engagement, however small, can hint at future paths toward addressing perhaps the most pressing political problem of the twenty-first century—the democratic deficit of global governance.

The ubiquity and centrality of ICT in contemporary life is beyond dispute. There is growing recognition that IR scholars need to refocus their theoretical lenses accordingly if they are to provide comprehensive accounts of world politics. However, the role of ICT evades easy categorization because it reflects the dialectical tensions that are inherent to globalization. ICT emancipate by enabling new forms of political participation, but it is also characterized by disturbing patterns of social exclusion and repression. Public sphere theory is a useful method for systemizing thought about these issues. But transnational publics are still an unfamiliar theoretical proposition in IR. This book is an attempt to map out this ill-defined conceptual terrain. I conclude by offering some suggestions for the future direction of transnational public sphere research.

CHAPTER 2

Reconstructing Habermasian Public Sphere Theory

Haacke describes the contribution of Habermasian-influenced IR theorists as threefold. They are able to: "(1) reveal the possibilities for change immanent in social relations; (2) offer a compelling normative base for its critique; and (3) illustrate real-world instances of a reconceptualised praxis" (Haacke, 1996: 256). Hitherto, international critical theory has been preoccupied with challenging the positivist bias in IR, and reinterpreting concepts such as sovereignty and security. Apart from a few exceptions (e.g., Comor, 1994; Gill, 1995; Keohane and Nye, 1998; Krasner, 1991), the role of ICT in world politics has been neglected. Although similar claims can be made of other schools of IR thought, this lack of analysis is perhaps most notable within the critical theory canon, as the media was a central theme of the Frankfurt school (e.g., Adorno and Horkheimer, 1979; Adorno, 1991). Public sphere theory is an ideal method for developing critical thought on these issues, and if designed adequately, it can correspond to all of the criteria that Haacke outlines above. It has the potential to contribute to the furtherance of critical research agendas in IR. For example, it has relevance to debates about the evolution of the state, discourse ethics, the expansion of moral and political community, contemporary forms of exclusion, and the transformative role of counterhegemonic social movements (e.g., Cox, 1999; Hoffman, 1991; Linklater, 1990c, 1994, 1996b, 1998; Neufeld, 1995).

This chapter reviews conventional public sphere theory in order to establish the theoretical basis for an investigation into transnational public spheres. In the first section, I revisit Habermas' classic expression of the concept in *The Structural Transformation of the Public Sphere* (hereafter referred to as *Structural Transformation* or *STPS*). I argue that public sphere theory still retains some purchase in aiding conceptualization of transnational deliberation, but that it must be rearticulated in terms of contemporary global sociopolitical conditions. Hence, Habermasian theory is distilled into its basic elements in order to establish the essential preconditions for a nascent public sphere. Three institutional foundations are identified, ability to communicate, separation from public authority, and adherence to the norms of publicity. The latter condition requires a sufficient degree of affinity between participants to engage in normatively structured discourse, which in the Habermasian public sphere is based on national citizenship.

I then turn to consider a number of difficulties with the Habermasian definition of the public sphere that need to be resolved before the concept can be appropriated. Examples include the dichotomous division between public and private realms, nonrecognition of counterpublics, and overidealization of the bourgeois public sphere. Particularly problematic is the way in which the public sphere is conceptualized as territorially delimited. Even radical critiques of Habermas' work have failed to problematize this underlying state-centricity (e.g., Fraser, 1992; Ryan, 1992). The nation-state has been conventionally understood to be a sovereign power and therefore the obvious addressee of public deliberation. This assumption may have seemed reasonable before the onset of contemporary globalization. But classic theories of the public sphere originate from times substantially different from the present. Communication media had a more limited reach, and the nature of state sovereignty was less problematic (Baynes, 2001). An unavoidable question in contemporary public sphere theory is whether the growth of "global governance," coupled with the increased capacity to communicate across state borders, has contributed to further significant transformation of public spheres (Stevenson, 1993: 67).

The second section provides a brief summary of Habermas' recent writings, and discusses the revisions he has made to his approach. Indeed, he has foreshadowed many of the themes of this inquiry in his hypotheses on public deliberation in the "post-national constellation." Although Habermas evidently believes that public sphere theory has application in the present, opinions about its explanatory utility and normative worth are divided. For example, Jodi Dean maintains that the concept of transnational public spheres is essentially flawed and that the relationship between ICTs and international social movements are more productively conceptualized through the prism of "global civil society" (Dean, 2001). Therefore in the third section, with reference to Dean's relativist critique, I contend that a reformulated public sphere theory can offer a sophisticated analytical critical perspective and unique insights into world politics.

2.1 Habermas and the Public Sphere

Public sphere theory has a variety of different incarnations, including Arendt's conception of the agnostic public (Arendt, 1958; for a critique, see Benhabib, 1995) and Dewey's description of historical small town meetings in the US (Dewey, 1927). However, the principal authoritative source is widely regarded to be Jürgen Habermas' seminal work: *STPS* (1999). Originally published in 1962, it has since spawned countless imitations. More recently it has also prompted a barrage of criticism. Indeed, some of the critiques have been later echoed by Habermas himself (e.g., Habermas, 1992a). Nonetheless, *STPS* remains a classic referent point for public sphere theory because of its ability to inspire and orientate critical thought about issues of deliberative democracy. It is a flawed but compelling account, methodologically appealing but laden with factual flaws. Unlike his later, more abstract work, Habermas uses historical-institutional analysis to identify an ideal model of discourse against which actually existing conditions can be assessed. Similarly, I assert that it is not possible to evaluate the revolutionary potential of transnational discursive spaces without understanding the process whereby early-modern political dialogue evolved into higher forms of critical publicity. Habermas' model of the public sphere holds value for investigating the possible emergence of transnational public spheres, but requires significant adaptation. A rereading of Habermasian theory reveals a set of social conditions underpinning critical publicity that can be abstracted to analyze emancipatory possibilities in the present.

The thesis of *STPS* is well-known and does not require detailed reiteration here. A brief recount of the main tenets will suffice. Habermas describes a public sphere as a realm of free and open discussion, oriented toward consensus, where the merit of argument determines outcomes rather than the socioeconomic status of participants (Habermas, 1999: 27–30). Private utterances can be distinguished from public statements in that the latter are addressed to an indefinite audience, with the expectation of a response. Hence, public spheres are universally inclusive in principle as anyone can join the debate. In *political* public spheres, public opinion is formed about issues of governance and addressed to the sovereign power. In Habermas' words, it is

> ...a forum in which the private people, come together to form a public, readied themselves to compel public authority to legitimize itself before public opinion. The *publicum* developed into the public, the *subjectum* into the reasoning subject, the receiver of regulations from above into the ruling authorities' adversary. (25–26)

The notion encapsulates the very essence of democracy. These ideals have patently never been fully realized, but it is nonetheless possible to identify a variety of historical approximations. Habermas describes the emergence of a public sphere amongst the literate bourgeois in the *salons* and coffeehouses of eighteenth-century Europe (57–66). These venues can be thought of as something akin to the Ancient Greek agora, and served as latter-day public forums for private citizens to discuss affairs of state. This public sphere arose in the context of a budding urban culture, improved transportation and the emergent media of newsletters and journals. The result was greatly increased social intercourse, albeit only within a rarefied echelon of the bourgeois (31–43). Habermas characterized the public sphere as a realm of informed and reasoned debate, where government policies were scrutinized and arguments and opinions rationally discussed.

It is possible to identify "institutional criteria" relating to the public sphere according to Habermas' historical description of its emergence. First, the public sphere developed in separation from the state as a domain of citizen empowerment. The state bureaucracy represented domination, power, and public authority; the public sphere represented rationality, private autonomy, and the demand for public accountability. The public asserted itself as the source of legitimation for the exercise of political power:

> In this [bourgeois] stratum, which more than any other was affected *and* called upon by mercantilist policies, the state authorities evoked a resonance leading by the *publicum*, the abstract counterpart of public authority, into an awareness of itself as the latter's opponent, that is, as the public of the now emerging *public sphere of civil society.* For these latter developed to the extent to which the public concern regarding the private sphere of civil society was no longer confined to the authorities but was considered by the subjects as one that was properly theirs. (23, original emphasis)

Second, communication media played an integral role in the development of the public sphere:

> Because...the society now confronting the state clearly separated a private domain from public authority and...a subject of public interest, that zone of continuous administrative contact became "critical" in the sense that it provoked the critical judgment of a public making use of its reason. The public could take on this challenge all the better as it required...the function of an instrument with whose help the

state administration has already turned society into a public affair in a specific sense—the press. (24)

Third, members in the public sphere recognized the moral validity of political debate that was maximally inclusive of all actors affected by the exercise of sovereign power. In other words, the demos constituted the public sphere in principle (even though the public was exclusionary in practice). This required participants to identify with the social imaginary of the nation-state as a "public." Habermas argues that even where numbers were small, gatherings were *conscious* of themselves as representative of a wider public. In Habermas' words:

> Wherever the public established itself institutionally as a stable group of discussants, it did not equate itself with *the* public but at most claimed to act as its mouthpiece, in its name, perhaps even as its educator—the new form of bourgeois representation. (37)

This sense of affinity was exemplified by the norms of publicity that were embodied in discourse. Participants demonstrated an expectation of dialogue by putting forth intelligible opinions and arguments for which they were prepared to be held to account. Further, they exhibited a commitment to rationality by excluding direct appeals to power as a means of settling disputes. Disparities of wealth and status were transcended in the bourgeois sphere in favor of "the authority of the better argument [which] could assert itself against that of social hierarchy and in the end carry the day" (36). The critical dialogue thus generated was oriented toward understanding. Therefore, the basic preconditions of the emergence of the public sphere can be summarized as follows: separation from public authority (the state acting as the addressee of public sphere debate), ability to communicate (via the medium of print), and adherence to the norms of publicity (which requires a sufficient degree of affinity between participants to engage in normatively structured discourse).

For Habermas, the emancipatory significance of the public sphere was immense. It represented a fundamental change in the organization of society, from feudal rule to the public use of reason to arbitrate between competing power claims. Thus, it encompassed a normative transformation in the nature of the power claims. The bourgeois public began to articulate their interests in ways that "would entail, if it were to prevail, more than just an exchange of the basis of legitimation while domination was maintained in principle" (28). Inherent in the bourgeois conception of democratic legitimacy was the emancipatory ideal of balancing state power with public opinion. Thus the political power of the public sphere lay in its potential to neutralize the state as a tool of domination (27).

Habermas' account of the degradation of the public is almost unremittingly pessimistic. The dramatic arc of decline seems especially stark in contrast to the apotheosized bourgeois era. He recounts how the potential for critical discourse was radically curtailed by the triumph of corporate capitalism, the manipulation of popular opinion by the advertising industry, and the rise of a passive consumption mentality amongst the masses (181–235). Habermas conceives the rise of the public sphere as dependant on a clear separation between public and private interests, and so argues that the gradual integration of both was corrosive. It produced an intermediate sphere of state actors absorbed in society, and social actors absorbed in the state. Critical publicity was not engendered by private people in the intermediate sphere. Instead

> The process of the politically relevant exercise and equilibration of power now takes place directly between the private bureaucracies, special-interest associations, parties, and public administrations. The public as such is included only sporadically in this circuit of power, and even then is brought in only to contribute to its acclamation. (176)

An effective illustration of the decline of the public sphere is the semantic change of the term "publicity." Historically, "publicity" has been understood to denote the condition of being public; but presently, it is more commonly associated with manipulative political tactics. Likewise, "public opinion" does not suggest active critical participation as much as it once did in the early-modern era. Rather, public opinion is often seen as a prize that politicians and advertisers struggle to capture and shape for their own political and economic interests. As Peters (1993) points out, the changing subtextual connotations of "public" and "private" contributes to the structural transformation of the public sphere.

For Habermas, two tendencies arise from the emergence of the intermediate sphere and the social welfare state. There is an antidemocratic tendency, where organizations compete with one another to maximize their influence over the policy process chiefly by securing consensus through manipulative publicity (Habermas, 1999: 178). Thus public debate is distorted and public opinion is manufactured by powerful sectional interests. However, there is also an ameliorative democratic tendency, insofar as the state enshrines the constitutional rights of all citizens (232–233). Habermas argues that the bourgeois version of the public sphere of private people cannot be effectively replicated in this context. Instead, the public sphere is recast as dialogue between social intermediaries. With a tone of despondent resignation, this is where Habermas locates immanent possibilities for the democratization of contemporary Western nation-states:

> *Only such a public could under today's conditions, participate effectively in a process of public communication via the channels of the public spheres internal to parties and special-interest associations and on the basis of an affirmation of publicity as regards the negotiations of organizations with the state and one another.* (232, original emphasis)

Structural Transformation has retained significant influence decades after it was written because it systematically addresses a central problem in normative theory: what are the conditions for linking public opinion to the mechanisms of governance, so that rationality can be privileged over power claims? As Boyd-Barrett argues, the impact that *STPS* continues to exert also relates to

> the weight it gives to the everyday culture of a social class and its use of the media gives it a sociological, not to say ethnographic, authenticity which is impressive and which underlines the dearth of equivalent work for other media in other historical and social contexts. (Boyd-Barrett, 1995: 231)

The strengths of the Habermasian approach emanate from the critical theory tradition in which it is anchored. Public sphere theory is oriented by an emancipatory interest and a critical stance in relation to existing social conditions. It is informed by a visceral commitment to the notion that governance only derives legitimacy from the common consent and active participation of the people.

This standpoint has attracted vitriolic criticism from some postmodernists, who attack the "metanarratives" of rationality and universalism in Habermas' work (e.g., Lyotard, 1984: 65). Mark Poster succinctly describes the sources of postmodernist unease:

> …when Habermas defends with the label of reason what he admires in Western culture, he universalizes the particular, grounds the conditional, absolutizes the finite. He provides a centre and an origin for a set of discursive practices. He undermines critique in the name of critique by privileging a locus of theory (reason) that far too closely resembles society's official discourse. (Poster, 1989: 23)

However, critique that proceeds from a postmodern rejection of universal norms is restricted to a discourse of unconstructive resistance (Blaug, 1994). This detatchment is wholly unsatisfying in the context of endemic global poverty, injustice, and inequality. In contrast, critical theory is oriented by an emancipatory interest in transformation. As Calhoun argues, to be meaningful, politically and theoretically, theory must provide a base that allows "for critical judgements to be arguable, defensible, in discourse across the lines of cultural, ideological, or other differences" (Calhoun, 1995: 119). Radical relativists do not provide such a base. In contrast, Habermas' advocacy of a maximally inclusive public sphere underpins his critique of the status quo and frames his proposals for alternative futures.

Nonetheless, if reflexivity is a core characteristic of a sophisticated critical theory, then dialogue with the postmodernists warrants serious engagement by Habermasian theorists. Postmodern critiques often flag up epistemological issues that require careful consideration. As his contribution to Calhoun's 1992 volume indicates, Habermas regularly invites external critique of his work, indulges in self-analysis, and reevaluates the validity of his theoretical assumptions (e.g., Habermas, 1992a). In the same spirit, it is essential to acknowledge that the dangers inherent in universalist perspectives should not be ignored or underestimated. Universalism can contain an instrumentalist tendency for control and domination. The question is whether these dangers can be adequately mitigated by a reflexive standpoint and a minimalist universalist approach.

I maintain that this is possible. Public sphere theory is undeniably foundationalist, but it eschews some of the inherent hazards of universalism because it concerns only procedural matters. It focuses on the process by which decisions are taken—it is interested in the social conditions that permit dialogue to be as open, free, and fair as possible. Public sphere theory espouses certain conditions for discourse, but does not prescribe substantive outcomes to debate. This would divest citizens of their right to speak, which is directly counter to the public sphere's raison d'être. Indeed, individual expression is strongly defended by public sphere theory (Benhabib, 1992: 84). In this way, Habermas persuasively balances universalism with respect for difference (Holub, 1991: 161).

However, there are serious difficulties with Habermas' approach that have received persistent and justified criticism over the years (some examples from a massive literature include Curran, 1991; Mahieu, 1988; Scannell, 1989; Schudson, 1992). These limitations need attention before it is possible to consider how the concept of a public sphere can be adapted for analysis at the transnational level. Therefore, a critique of Habermasian theory follows, divided into five thematic areas that cover the most problematic aspects of Habermas' thesis: alternative histories of the public sphere, social equality and democracy, the singular public sphere model, private interests, the potential of new media, and state-centricity.

2.1.1   On Alternative Histories of the Public Sphere

It has been a long-standing criticism of Habermas that he idealizes the bourgeois public sphere and exaggerates the extent of free debate (Schudson, 1992). Postmodernists have argued that Habermas has constructed an idealized past to fit an underlying "grand narrative" (e.g., Poster, 1989). It is now commonly accepted that a public sphere as described in STPS has never been realized, "at best there has been some initiative" such as acceptance of the right to free speech (Verstraeten, 1996: 349). In fact, the bourgeois public was defined by key exclusions of gender (Fraser, 1992) and class (Eley, 1992). Indeed, Habermas has since conceded this point (Habermas, 1992a: 463). Structural Transformation is fundamentally problematic as it rests on the proposition of a public sphere that is based on historical inaccuracies and misrepresentations.

Habermas' tendency toward overstatement is exemplified by his differential treatment of the bourgeois public sphere and the transformed public sphere of late capitalism. As Calhoun argues

> Habermas tends to judge the eighteenth century by Locke and Kant, the nineteenth century by Marx and Mill, and the twentieth century by the typical suburban television viewer. Thus Habermas' account of the twentieth century does not include the sort of intellectual history, the attempt to take leading thinkers seriously and recover the truth from their ideologically distorted writings, that is characteristic of his approach to seventeenth, eighteenth and nineteenth centuries. (Calhoun, 1992: 33)

Moreover, his coverage of the earlier period neglects the popularity of precursors to the tabloid press, such as "scandal-sheets," and the rabble-rousing of public agitators (ibid.). Such examples suggest that Habermas overestimates the zenith of the public sphere and its subsequent degradation.

Further, Habermas does not recognize the emergence of public spheres that developed in tandem with, and often in direct conflict with, the liberal bourgeois version. For instance, Ryan (1992) documents the variety of ways in which nineteenth-century North American women of different classes and races accessed the official public sphere when suffrage only extended to males. Her study ranges from politically active groups of elite women to working-class women in trade unions. Habermas also pays little attention to the contemporaneous nationalist, peasant, and proletariat publics (e.g., Warner, 1992, 2002; Negt and Kluge, 1993). These counterpublics challenged the exclusionary boundaries of the bourgeois sphere and promoted oppositional discourses and values. For example, the solidaristic norms espoused by counterpublics contested the individualist ethos of capitalism (Keane, 1984: 29). Habermas treats these exclusions and conflicts as merely characteristic of the period, but they deserve more serious consideration. To neglect such evidence is indicative of a bourgeois, masculinist bias (Meehan, 1995; Young, 1996). There is a need for a keener critique of the official public sphere and a greater awareness and sensitivity about the contributions of different sections of society to political life.

2.1.2   On Social Equality and Democracy

Habermas describes the bourgeois public sphere as accessible to all in principle, even though this was never fully realized in practice. There were entrenched exclusions on the basis of gender, race, and property ownership. Habermas appears to suggest that social inequalities do not affect the normative value of the public sphere as dialogue is premised on equality of moral-political status. For example, with regard to Carol Pateman's skepticism about whether women could be integrated into a patriarchal public sphere (Pateman, 1988: 82), Habermas argues that "this convincing consideration does not dismiss rights to unrestricted inclusion and equality, which are an integral part of the liberal public sphere's self-interpretation, but rather appeals to them" (Habermas, 1992a: 429). Habermas propounds a more inclusive version of the public sphere that abides by the norms upheld in the bourgeois manifestation (Habermas, 1999: 36). His argument is anchored on an assumption that he does not fully explore—that the norms of publicity can be realized in the context of gross social inequality.

Revisionist historical research suggests that social inequalities structured the debate in the bourgeois sphere. As Nancy Fraser observes, "discursive interaction within the bourgeois public sphere was governed by protocols of style and decorum that were in themselves correlates and markers of status inequality" (Fraser, 1992: 119). Political deliberation can be conducted to exert subtle forms of domination. Thus in theorizing about the public sphere, one should be careful of the rhetoric of the "common good," and sensitive to the unequal power relations it may uphold. Deliberation framed from such a standpoint may serve to validate the hegemonic ideology of dominant groups and delegitimize the grievances of the marginalized. These effects are reinforced by the subtextual connotations of language. Subordinate groups may find it difficult to articulate their thoughts and demands, or to do so in ways that ensure that they are heard and taken seriously. Their interests are often ignored or misinterpreted.

Habermas' argument presumes that rational deliberation in the public sphere can eclipse cultural differences. But all societies privilege certain cultural styles other others, which effectively devalues the contributions of subordinate groups (Young, 1996: 123–124). For instance, it has historically been the case in the West that the educated white voice has been accorded greater legitimacy than that of the illiterate black. The Habermasian public sphere promoted a version of the public interest that was authored by bourgeois males and suppressed alternative voices. A rich legacy of feminist research has revealed the multiple ways in which women are silenced and discouraged to express their needs. It is within the feminist counterpublic that women have found the freedom and deliberative spaces to articulate their experiences of gendered oppression (Landes, 1988, 1998; Meehan, 1995). The homosexual, ethnic minority, and proletariat counterpublics have performed similar roles (e.g., Negt and Kluge, 1993; Norton, 1992).

Counterpublics emerge as a reaction to cultural exclusions to full participation in the dominant public. Economic inequalities usually mirror this cultural divergence, forming a further barrier to access for subordinate groups. Therefore, despite universal legal entitlement to participation in public life, exclusionary outcomes are magnified by cumulative inequalities. Fraser argues that these considerations raise the question of whether effective deliberation can ever be achieved in the face of deep-seated social inequality (Fraser, 1992: 119). The universal public sphere may be a fundamentally unfeasible prospect. This has been a common theme of critical theory, and one that deserves more sophisticated treatment than it receives in *STPS*. It leads to the next substantive criticism.

2.1.3   On the Singular Public Sphere Model

Habermas portrays the public sphere as sui generis and as an overarching, comprehensive entity. Not only is this factually erroneous but it is also informed by a presumption that democracy is best served by a single public sphere. In the absence of specialized arenas for dialogue, such a sphere is likely to put subordinate groups to further disadvantage (Keane, 1984: 29). It would mean they would have to deliberate under the "surveillance" of the dominant groups, which would render them less likely to articulate and defend their common interests. As discussed above, the prevailing discourse would also be framed according to a conception of the "common good" that was likely to reflect the interests of the powerful.

Again, revisionist historical research is insightful here. It reveals that marginalized groups have consistently sought out deliberative arenas distinct from the bourgeois sphere to challenge the hegemonic ideology embodied in mainstream discourse. Women, the working classes, ethnic minorities, and homosexuals relied on these counter-publics to develop understandings about their identities and interests that were not able to be fully explored in the dominant public sphere (Eley, 1992: 306; Fraser, 1992: 116; McLaughlin, 1993; Norton, 1992). It was a direct reaction to formal and informal mechanisms of exclusion in conventional political life. The counterpublics contributed to the expansion of discursive space (Warner, 2002). They were creative havens where new concepts were generated to describe experiences of prejudice and subordination (e.g., "sexism," "homophobia"). Of course, contemporary examples of counterpublics are abundant. In a large-scale, diverse society, the promotion of a singular public sphere is at best misguided and utopian, at worst sinister and oppressive.

Some critics perceive counterpublics as having a deleterious effect on democracy, arguing that they are insulated from wider society and fragment national debate. But this is not the case if discourse is characterized by the norms of publicity. Members of counterpublics aspire to secure greater recognition and inclusion within the dominant public. The new language of marginalization is introduced into wider political sphere in order to reframe the hegemonic discourse and to press the case for the reform of societal institutions. In this respect, counterpublics are emancipatory (Fraser, 1992: 115–116). Participatory parity is best served by a multiplicity of publics, and historically this has always been the case.

2.1.4   On Private Interests and the Public Sphere

Habermas theorizes the public sphere as a place where it is possible to divorce oneself from private interests and engage in reasoned debate oriented toward consensus (Habermas, 1999: 24). This conception draws on the conventional patriarchal definition of the division between the public and private realms (Calhoun, 1992: 35–36). For Habermas, there is an unambiguous distinction between public and private issues and it is always undesirable for the latter to intrude upon the former. It is argued that public opinion would otherwise be distorted in favor of sectional interests.

However, the public/private dichotomy is a powerful ideological tool to foreclose dialogue about the oppression of women (Pateman, 1987). Women have been traditionally relegated to the private sphere and thus denied an effective political voice. To revive an old cliché, for women, the personal is political. It is within the feminist counterpublic that women have generated discourses and new language to describe their experiences of repression, such as domestic violence and marital rape (Landes, 1998; McLaughlin, 1993). In the bourgeois public sphere, the conspiracy of silence about such "domestic" matters benefited the male perpetrators of abuse. Therefore, although Habermas professes to have a normative interest in a universally accessible public sphere, he actually legitimizes conditions of political deliberation that reify gendered exclusions (Benhabib, 1992: 93).

The public/private divide does not only disadvantage women. By defining the economy as part of the private sphere, Habermas also excludes issues regarding capitalism and class. The grievances of the proletariat are silenced to the profit of the bourgeois class (Negt and Kluge, 1993). It is important to remember that rhetoric regarding the "common good" can mask multiple exploitative social relations. It often serves to invalidate certain viewpoints, topics and interests and prioritize those of the most powerful (Calhoun, 1992: 37). There is not a pre-given boundary that demarcates "private" from "public" matters. Neither is there a neutral way of distinguishing between them. It is imperative in public sphere theory to reassess the theoretical constructions that are implicit in the reproduction of domination.

Public sphere participants will disagree about what subjects are appropriate for debate. Therefore no area of life should be axiomatically ring-fenced as taboo (Benhabib, 1992: 93). Rather, if dialogue is to be truly inclusive, marginalized groups must be free to argue that issues commonly perceived as private should be brought into the public domain (Phillips, 1998). This remains an area of difficulty with Habermas as he regards the maintenance of the division as important, and it is rearticulated in his thesis of systemic intrusion of the lifeworld (e.g., Habermas, 1987).

2.1.5 On the Potential of New Media

The extent of Habermas' pessimism about the decline of the public sphere is unwarranted, as it relates to an overidealized conception of early-modern discourse (Thompson, 1990: 115–121, 1995: 73–75; Keane, 1984). Drawing perhaps too much from the cynicism of Adorno's cultural industries model[1], Habermas overstates the ability of contemporary media to promote political apathy (Garnham, 1992: 360). In contrast, some theorists argue that ICT

> greatly increases the visibility of political leaders, and limits the extent to which they can control the conditions of reception of messages and the way in which these messages are interpreted by recipients…leaders [have] a new visibility and vulnerability before audiences which are more extensive and endowed with information and more power (however intermittently expressed) than ever before. (Thompson, 1990: 115)

In addition, this technological infrastructure facilitates the emergence and mobilization of transnational social movements, which have become increasingly significant political actors in recent years (Adams, 1996: 419). In this context, ICT is an agent of democratization. It can enhance transparency and accountability, and help to recast the power balance between citizens and decision-makers.

It could be argued in Habermas' defense that he could not be expected to foresee the full implications of developments in ICT at the time of writing *STPS*. However, excessive pessimism is evident within his most basic assumptions. For example, Habermas' analysis of the media's influence at a meta-institutional level obviates the "micro-level" of the individual's interpretations to media stimuli (Dahlgren and Sparks, 1991: 10–21). The citizen cannot be reduced to little more than a passive consumer of messages from "the top." Media content is filtered through the complex intersubjective understandings of the audience, and so the audience also has an active role in "producing" meanings. Of course, the scope of this is circumscribed owing to the asymmetry of the relationship between the creator and receiver of messages (Verstraeten, 1996: 350). However, the public sphere is only realizable through a symbiotic relationship between the media and the people, and the potential for resistance and empowerment is located in the audience's processes of active signification. The challenge for contemporary public sphere research is to explore how profound shifts in media structures and technologies affect these emancipatory possibilities. The corrosive pessimism of

Habermasian analysis forecloses such inquiry. But it presents a bleak outlook for democratic deliberation that may not be justified.

2.1.6   On the State-centricity of
Conventional Public Sphere Theory

The bourgeois conception of the public sphere is embedded in the nation-state. The state apparatus is a crucial institutional precondition; it implicitly defines the boundaries of the public in terms of membership and the circulation of discourse (Habermas, 1999: 82). The national government is the recipient of public demands for action and accountability. In Habermas' words: "The constitutional state as a bourgeois state established the public sphere in the political realm as an organ of the state so as to ensure institutionally the connection between law and public opinion" (81). This version of public sphere theory implicitly assumes equivalence between the virtual space of public deliberation and the physical space of the nation-state. However, these background presumptions need to be reassessed if discursive spaces and sites of governance have expanded in recent years.

Conventional public sphere theory was designed to raise questions regarding the implications of corporate media ownership for the rationalization of governance (Calhoun, 1992: 41). But these issues are not necessarily exclusive to the domestic realm. *Structural Transformation* should be read in terms of the author's intention to contribute to normative critique of the democratic shortcomings of modern nation-states (Fraser, 2005). Habermas is concerned with the sovereign domination of national governments, and the manipulations of the popular press. He was writing at a time where politics and media were commonly conceived in terms of bounded territorial spaces. Framing the public sphere in the context of the nation-state tacitly relies on three main presuppositions. First, there is a national communications network, including mass media that report affairs of government (Habermas, 1999: 73). Second, sovereignty is coterminous with political territory (74). And third, that members of the public sphere have common interests by virtue of their national citizenship (83).[2]

Contemporary technological, political, and social trends suggest that all three assumptions need to be thoroughly problematized. These include the rise of new media, the increasing multifaceted challenges to state authority associated with "global governance," and the growth of cross-border social movements that articulate an alternative basis for shared identity. These developments may have historical precedents. For example, Deibert (1997) traces the evolution of international communication through the centuries; Hirst and Thompson (1996) dispute the "novelty" of many attributes of "globalization," and Keck and Sikkink (1998: 39–78) describe the antislavery and suffrage movements as early examples of transnational advocacy networks.

Nevertheless, in the present era these trends are of increased salience, and at the very least, require that any recasting of public sphere theory must seriously question the validity of the nation-state paradigm.

Statist assumptions have persisted even in radical critiques of Habermas' work. The social exclusions of the bourgeois public have been documented in great detail, but usually only in terms of how these were manifest within the nation-state (e.g., Fraser, 1992; Ryan, 1992; Meehan, 1995). The extraterritorial dimension has not been adequately theorized. Indeed, in a perceptive reappraisal, Fraser has recently acknowledged the state-centricity of her earlier work. She explains that the statist presuppositions of bourgeois and radical models of the public sphere derive from the similar normative orientation of both. In her words,

> ... classical public sphere theory constituted a critical theory of a specific political project: the project of modern Westphalian-national state democratization. The critique of this theory has focused largely on securing the full inclusion of those nationals who were excluded or marginalized within that frame: propertyless workers, women, racial minorities, and the poor. (Fraser, 2005)

The bourgeois incarnation and other national variants are transitional embodiments of the public sphere. Critical publicity is possible in other contexts; it is not exclusive to the nation-state. A public sphere describes a mode of discursive engagement in relation to political authority, rather than a bounded geographic space. It is erroneous to assume that a prerequisite for norms of publicity is geographical proximity *per se*; rather what is required is a correspondence between deliberative spaces and sites of authority. The tacit implication in treating the state as a pre-given is that the national political and media apparatus are immutable structures, which is evidently suspect.

Not only can these statist presuppositions be questioned on *prima facie* grounds, but also can be criticized from the normative standpoint of critical international theory. A state-centric conception of the public sphere restricts theoretical inquiry to a bounded territory, effectively foreclosing analyses into immanent possibilities for world order transformation. This is incompatible with a universal interest in human emancipation. Therefore, there are both evidential and normative reasons to challenge the state-centricity of Habermas' approach.

2.2 HABERMAS: RECENT WRITINGS

Subsequent to *STPS*, Habermas established a different orientation for critique through theories of language and communication (Cooke, 1994). His focus switched from sociohistorical and political-institutional analysis to a more abstract and philosophical inquiry about the innate emancipatory potential in language itself (see, among others, Habermas, 1984, 1987, 1992b). For Habermas, rationality inheres in "communicative action," in other words, in the ability to comprehend dialogue, to submit to a superior argument, and to reach consensus (Habermas, 1984: 286–287). The logic of communicative action contains norms for processes of discursive will-formation, and condemns distortions and domination in communication. Habermas posited quasi-transcendental grounds for social critique through a model of deliberative interaction, termed "the ideal speech situation" (88).

This "linguistic turn" precipitated profuse theoretical debate, which need not be reviewed here. My purpose in this inquiry is to reconstruct Habermasian public sphere theory, which does not encompass acceptance of Habermas' later foundational claims concerning human reason, communicative action, and argumentation. *Structural Transformation* is distinct from the rest of Habermas' oeuvre insofar as it provides an account of the public sphere that is located in historical conditions and is associated with a specific kind of sociability. Habermas' later communicative action theory is far more metaphysical. This investigation has been inspired by the situatedness of *STPS*. However, there are certain relevant continuities in Habermas' recent work with his early writings on the public sphere.

Habermas has conceded the validity of many of the aforementioned criticisms and sought to incorporate these into a reformulated theory of the public sphere (e.g., Habermas, 1992a). His cynical assessment of the decline of the bourgeois public seems to have been revised somewhat in favor of a more cautiously optimistic view that "under certain circumstances, civil society can acquire influence in the public sphere" (Habermas, 1992b: 373). However, he maintains that his analysis of the state-capital domination of the public sphere is still applicable (Habermas, 1992a: 441).

Habermas now prefers a more inclusive definition of the public, where "its institutional core comprises those nongovernmental and non-economic...voluntary associations that anchor the communications structures of the public sphere" (Habermas, 1992b: 366). In this conceptualization, the public sphere acts as a transmission belt where, via the intermediaries of civil society and the media, the concerns of the marginalized can reach the mainstream political agenda. Habermas sees the growing influence of feminist and environmental movements over the years as illustrative in this regard (ibid.). He also distinguishes between the "universal public sphere" and the "pluralistic, internally much differentiated mass public," which acknowledges the critique of the singular sphere model (Habermas, 1992a: 438). How does Habermas reconcile the resulting tension between universalism and respect for difference? He asserts that no public can permanently intend to exclude others: "there is no exclusion mechanism without a proviso for its abolishment" (Habermas, 1992b: 374). In other words, he believes that all public spheres must have the potential for self-transformation: "The labor movement and feminism were able to join these discourses in order to shatter the structures that had initially constituted them as 'the other' of a bourgeois public sphere" (ibid.). In effect, this fully incorporates Fraser's critique (see also Habermas, 1992a: 458; Habermas, 1996: 374).

Further, Habermas provides a "two-track" solution to the problem of cultural diversity and social heterogeneity of contemporary states, stating that mass participation in the public sphere must be connected to reasoned and competent political decision making (Habermas, 1992a: 452). The executive and legislature provides an institutional focus for deliberation within the informal public sphere. It is vital that these governing institutions should be responsive to public opinion if deliberation in the public sphere is to be truly meaningful. As Habermas contends

> the public sphere must...amplify the pressure of problems, that is, not only detect and identify problems but also convincingly and *influentially* thematize them, furnish them with possible solutions, and dramatize them in such a way that they are taken up with and dealt with by parliamentary complexes. (Habermas, 1996: 359, original emphasis)

Habermas reflects that it will often be more difficult for smaller public spheres to wield this type of mediated influence, but not impossible. The presentation of marginalized issues is key: "only through their controversial presentation in the media do such topics reach the larger public and subsequently gain a place on the 'public agenda'" (ibid.: 381). Thus norms of publicity inherent in the dominant, mass-mediated public can be exploited by subordinate groups as a "latent tendency" toward wider inclusion that may secure greater recognition of their concerns (ibid.).

Although the mass media are central in Habermas' modified conception of the public sphere, he remains ambivalent to the deliberative potential of new media. He warns that although technologies such as the Internet open up new channels of communication, they also promote the fragmentation of civil society:

> Whereas the growth of systems and networks multiplies possible contacts and exchanges of information, it does not lead per se to the expansion of an intersubjectively shared world and to the discursive interweaving of conceptions of relevance, themes, and contradictions from which political public spheres arise. The consciousness of planning, communicating, and acting subjects seems to have simultaneously expanded and fragmented. The publics produced by the Internet remain closed off from one another like global villages. For the present it remains unclear whether an expanding public consciousness, though centered in the lifeworld, nevertheless has the ability to span systematically differentiated contexts, or whether the systemic processes, having become independent, have long since severed their ties with all contexts produced by political communication. (Habermas, 1998b: 120–121)

In addition, Habermas has written extensively on the decline of the nation-state due to the combined challenges of multiculturalism and economic globalization, and argued that new democratic steering mechanisms must be developed in the emergent "post-national constellation" (Habermas, 2000). He sees the potential for new cosmopolitan solidarities to develop within transnational public forums. He considers environmental movements like Greenpeace as representative of an emergent cosmopolitan consciousness and nascent global citizenry. However, he is also concerned that broader forms of publicity should correspond with an investment in decision-making mechanisms on an international scale that can assume some of the political sovereignty and competencies of nation-states. In this regard, he has been particularly interested in the European Union (EU) as an experimental blueprint for a process of international cooperation and integration that could be paralleled elsewhere, with emphasis on the concept of constitutional solidarity (e.g., Habermas, 1998b: 105–127). The public sphere still remains the indelible motif of his thought,

although he maintains that the potential for a transnational European
sphere has not yet been realized:

> There will be no remedy for the legitimation deficit, however, without a
> European-wide public sphere—a network that gives citizens of all mem-
> ber states an equal opportunity to take part in an encompassing process
> of focused political communication…So far, however, the necessary
> infrastructure for a wide-ranging generation of diverse public opinions
> exists only within the confines of nation-states. (Habermas, 2001: 18)

The above review is evidently not a comprehensive account of
Habermas' post-*STPS* output, instead it serves to highlight those
themes of direct relevance to this inquiry. It also illustrates that *STPS*
is not a definitive statement; certainly Habermas would not claim it to
be so. Subsequently, he has made substantial revisions to his theory of
the public sphere, not least by acknowledging the existence of counterpublics and considering the implications of ICT and global governance on the continuing transformation of public spheres. Thus,
STPS best serves as a useful *entrée* to theorizing critical publicity.
Craig Calhoun best describes its legacy as "an immensely fruitful
generator of new research, analysis and theory" (Calhoun, 1992: 41).
Likewise, here it serves as a point of theoretical departure.

Elsewhere, the case has been put that a civil society perspective is a
more appropriate approach than the public sphere by which to analyze the implications of new media (Dean, 2001). The claim is important to consider carefully, especially in the light of the various problems
associated with the public sphere discussed above. I therefore turn to
evaluate the utility and validity of the contrasting frameworks.

2.3 GLOBAL CIVIL SOCIETY THEORY AS AN ALTERNATIVE APPROACH

There is a notable tendency in the literature to conflate the concepts
of transnational public spheres and global civil society. For example,
in their study of transnational advocacy networks, Keck and Sikkink
argue that "[t]he new networks have depended on the creation of a
new kind of global public (or civil society), which grew as a cultural
legacy of the 1960s" (Keck and Sikkink, 1998: 14). However, the
terms "civil society" and "public sphere" are not synonymous and
should not be used interchangeably. As Downey and Fenton argue,
"…it is crucial to keep them distinct and analyze the relationship
between social institutions and discourse" (Downey and Fenton,

2003: 190). As will become clear, the subtle semantic differences have important implications for the task of conceptualizing contemporary modes of deliberation.

The term "global civil society" is relatively new (Cohen and Rai, 2000; della Porta et al., 1999; Falk, 1998; Lipschutz, 1992; Price, 1998) although the role of non-state actors in world politics has long been an important topic in IR (e.g., Keohane and Nye, 1972). Whom and what constitutes civil society is an abiding matter of contention. Anheier et al. have the following functional definition: "a supranational sphere of social and political participation," distinct from the practices of governance and economy, but existing "above and beyond national, regional and local societies" (Anheier et al., 2001: 4). This interaction between governance and global civil society ranges from formalized and frequent to ad hoc and sporadic (Charnowitz, 1997). At one end of the spectrum, the UN, the EU, and the World Bank make regular attempts to engage with civil society actors across nation-states (Alger, 2002). At the other, the WTO has been routinely criticized by civil society for a chronic lack of public consultation.

Whilst the concepts of civil society and public sphere are distinct, they are nonetheless closely related: "the former standing for structures and the latter for shared meanings emerging through these structures" (Sassi, 2001: 100). Yet the public sphere is a more appropriate theoretical approach for the purposes of this inquiry for two main reasons. First, public sphere theory is premised on a deliberative ideal, where reasoned argument prevails over the social status of actors. The definition of the concept lays the framework for normative analysis of the conditions for critical publicity. It contains an inherent means to critique those types of social organization that have negative implications for democracy. "Civil society" does not contain a similar method of evaluation, and thus can be used to refer to progressive and reactionary movements. For example, "civil society" can encompass antiracist movements as well as the Ku Klux Klan (Chroust, 2000). Advocacy groups in civil society can pursue repressive goals that may be in direct conflict to the norms of publicity (Amoore and Langley, 2004). In addition, the internal constitution of groups in civil society can vary widely. Some are democratic and transparent, and the leadership is accountable to its members. Others are quite the reverse. In the above definition, Anheier et al. imply that civil society is a counterweight to sovereign domination by acting as a "buffer zone" between the state and the private sphere. But such dominance may be replicated if civil society groups are internally hierarchical and undemocratic (Drainville, 1998). Unequal power structures within and without civil society can be effectively critiqued through a public sphere approach.

The second reason is that public sphere analysis pivots on issues of communication and technology, which are central themes of this inquiry. Mediated communication is an essential precondition of public sphere manifestation. Public sphere theory focuses on how technological, political, and social structures can be most effectively structured to generate critical publicity. In contrast, the concept of civil society does not rest on a notion of inclusion or contain a theory of communication and participation. Instead, civil society theory concentrates on the influence exercised by non-state actors on governments and decision-making mechanisms (for more on this theme, see section 6.1).

I want to explore this line of argument further by considering the opposing viewpoint of Jodi Dean, who critiques the notion of transnational public spheres, and argues that the civil society model is more appropriate for theorizing about the "transnational technoculture." This is because "the regulatory fiction of the public sphere privileges a theorization of political norms. Struggles that contest, resist, or reject its idealizations are excluded from the political terrain as remnants of tradition, say, or manifestations of a terroristic irrationalism" (Dean, 2001: 247). Adopting an essentially relativist stance, Dean argues that "politics is about unequal exchanges among people who have fundamentally different ways of reasoning, who have differing conceptions of what is normal and what is appropriate" (265). She critiques public sphere theory since it "limits the political to rational conversation among people who respect each other as equals" (ibid.). Dean posits that that a civil society approach is preferable because it is able to accommodate all discourse; further it "privileges the concrete institutions in which the subjects of politics come to practice, mediate, and represent their actions as political" (ibid.).

Dean's warnings about the tendencies toward domination in foundationalist discourses are pertinent. However, Dean's relativistic civil society approach infers that equal validity can be accorded to all forms of discourse, despite Dean's caveat that "normative implications" can still be drawn from certain practices (264). It is unclear how normative critique can be operationalized in the absence of explicit criteria to evaluate forms of discourse. The public sphere is a powerful representation of the norms of equality, inclusion and mutual respect, and so it is a useful heuristic tool for problematizing inequities and exclusions. The concept of civil society does not incorporate these norms in such a robust way, and so opportunities for critique can be lost. Nevertheless, Dean's warning that public sphere theory can fail to capture the rich diversity of transnational discourse should be heeded. This problem is significant for those public sphere theorists who hypothesize a single, unified public, such as Garnham (1992: 371). The difficulty can be circumvented by adopting a multiple spheres model akin to Nancy Fraser's conceptualization of "subaltern counter publics"—which refers to those deliberative associations of marginalized groups that are neglected in Habermasian theory (Fraser, 1992). Dean argues that she is

> not convinced that adding an *s* solves the problem of the public sphere...despite its best intentions, the multi-spheres approach reinforces the priority of a bourgeois or official public sphere as a goal site, as an ideal, as the fundamental arbiter of inclusion. (Dean, 2001: 248–249)

In fact, Fraser's multiple spheres model was conceived as a direct critique of Habermas' privileging of the bourgeois public. I do not read Fraser's approach as reification of the Habermasian perspective but rather as a successful theoretical challenge to the primacy of the dominant public through revisionist historiography.

Dean makes a further criticism about public sphere theory, which she argues "remains tied to and dependent on the state, reinforcing a state-centric conception of democracy" (ibid.: 249–250). Dean is correct that the background presumptions of public sphere theory have traditionally reflected a statist bias. This limitation can be addressed by reformulating the concept to incorporate the transnational dimension whilst retaining the valuable potential for normative critique. In the next chapter, I suggest a way in which this can be done.

2.4 Conclusion

As Hill and Montag observe, "...the term 'public sphere,' as Habermas introduced it, has today become an ultimately foundational (and therefore underinterrogated) concept..." (Hill and Montag, 2000: 2). In a globalizing era, it appears to be a concept in need of a significant overhaul. It is necessary to reconstruct public sphere theory from its basic presuppositions to develop a more reflexive and critical approach. Crucially, this reformulation must permit critical inquiry into the possibility of public spheres emerging beyond the nation-state. From the above appraisal of Habermasian theory, it is possible to glean a number of appropriate guidelines for conducting research into trans-national public spheres. First, a multiple spheres template is preferable to a singular sphere model. Second, research should be sensitive to the implications of unequal power relations on the extent of access and participation in public spheres. Third, the conventional patriar-chal division between issues relating to the "public" and the "private" realm is repressive, exclusionary and therefore invalid. Fourth, research should investigate the emancipatory potential of ICT such as the Internet. And finally, it is vital to reassess the state-centricity that is characteristic of conventional public sphere theory. Several scholars have already sought to apply public sphere theory to an international level, but as shall be seen, not all have consistently followed these precepts.

CHAPTER 3

Contending Theories of Transnational Public Spheres: Propositions for an Alternative Analytical Framework

Critical international theorists such as Edward Comor and Stephen Gill have convincingly argued that a theoretical approach underpinned by a critical epistemology can aid in the explanation and understanding of the sociopolitical implications of ICT (Comor, 1994; Gill, 1995). However, they have not developed a comprehensive research methodology. Nor have they explicitly considered the utility of public sphere theory in conceptualizing these issues, despite the evident thematic relevance. Critical-theoretic issues such as the moral-political validity of law, the legitimacy of governance, and the extension of dialogic participation are central to the public sphere perspective. Indeed, it is possible to argue, as Nancy Fraser does, that "something like Habermas' idea of the public sphere is indispensable to social theory" (Fraser, 1992: 111).

A body of scholarship has grown in recent years concerning the expansion of public spheres beyond the nation-state, which has shaped the contours of current thinking about critical publicity. Much of this commentary derives from Communication Studies and Sociology, where global governance issues tend not to receive sustained attention (e.g., Sparks, 2001). This chapter discusses the emerging IR literature, and identifies common strengths and shortcomings, thus paving the way toward the formulation of a synthesized approach. The argument proceeds as follows.

The first section encompasses a review of prior instances of engagement between IR and public sphere theory. Scholars have introduced different versions of public sphere theory to the discipline, each employing distinctive terminology (e.g., "global," "cosmopolitan," "international," and "virtual" sphere/s). For the sake of convenience,

I use "extraterritorial public spheres" as a generic term for all of these approaches. Debate between these scholars revolves around the membership, scope and deliberative potential of transnational public spheres. Some of the literature is underpinned by a misplaced assumption that the concept of a transnational public sphere is commonsensical and unproblematic. Other key works have attempted to specify the conditions of possibility for extraterritorial publics, and there are several themes of divergence. These involve disagreements over the definition of extraterritorial publics, whether extraterritorial publics should be defined as singular or multiple, the role of global governance frameworks, the relevance of global inequalities, and the extent to which cross-border deliberation can be emancipatory. The theoretical implications of each controversy are discussed, and the possibility of synthesis is explored by focusing on the work of James Bohman.

The second section corrals these insights to construct a functional definition of transnational public spheres, which attempts to preserve the normative force of the Habermasian formulation, and avoid the shortcomings of some recent literature. This conceptual model is the cornerstone for the subsequent investigation into the possible revitalization of the public sphere.

3.1 EXTRATERRITORIAL PUBLIC SPHERES LITERATURE

Post-positivist IR scholars have developed several leitmotifs that are relevant to this inquiry. There is a long-running debate in international critical theory about the tensions between claims of universality and difference, and how these may be addressed (Brown, 1992; George, 1994; Haacke, 1996; Hoffman, 1987). For example, Kratochwil (1989) has examined the role of public justifications in the development of norms. There is interest in the concept of an emergent global civil society (Charnowitz, 1997; Cohen and Rai, 2000; della Porta et al., 1999; Falk, 1998; Lipschutz, 1992). Andrew Linklater combines these concerns by calling for an extension of political community in the "post-Westphalian" era, when the territorial link between the nation-state and the public is severed (Linklater, 1998). Thematically, public sphere theory fits in well with this critical research program.

It is perhaps not surprising then that transnational public spheres receive the occasional casual reference in recent literature, as if they are actually existing institutions. For example, Dryzek asserts that: "Civil society and public spheres...exist in the international system" (Dryzek, 2000: 130). Likewise, Craig Calhoun claims that "an international public sphere clearly already exists" (Calhoun, 2003: 229). Alexander Wendt has recently proposed that: "relevant to the constitution of a collective steering agency at the international level are "transnational public spheres" trying to keep states democratically accountable" (Wendt, 2001: 213). In an explicit attempt to apply a communicative action framework to IR, Risse contends that: "The existence of a public sphere ensures that actors have to regularly and routinely explain and justify their behaviour...[and] they vary dramatically in international relations" (Risse, 2000: 21). The impression conveyed is that extraterritorial public spheres are so eminently plausible there is little profit to be had from extended reflection on the conditions for their existence. Such "commonsensical" terms of reference are unwarranted. Too often, the conceptual viability of cross-border publicity fails to be queried with sufficient analytical rigor. As a result, extraterritorial public spheres are theoretically and empirically ambiguous. Indeed, Calhoun makes a similar point when he muses on the need to "do real research to help replace the contest of anecdotes and speculations with reasoned debate in the public sphere" (Calhoun, 2003: 249).

It is certainly inadequate to treat cross-border publicity as inherent to media proliferation, despite the ostensible logic of equating the two. Complex problems are implicated in the task of recasting the public sphere in an internationally anarchic environment. Fraser observes that much public sphere theory has tacitly assumed a nation-state frame; indeed the very concept of "the public" is imbued with statist presuppositions (Fraser, 2005). But there is no common trans-border political citizenry, and how far international institutions can be considered as analogous to the nation-state is a matter of intense controversy. It is far from clear that public spheres can easily transmute to the transnational level. Furthermore, treating extraterritorial public spheres as a pre-given vitiates a full commitment to the values embodied by the public sphere. As Fraser forcefully argues, public sphere theory was not just developed to describe communicative flows; it was designed to contribute to a normative theory of democracy (ibid.). It represents an attempt to secure some degree of moral-political validity for public opinion; therefore issues of inclusion should be paramount. Without an explicit normative orientation, public sphere theory risks bring denuded of its capacity for effective social critique.

Other international critical theorists have also recognized a pressing need for further research into the institutional foundations for extraterritorial public spheres (Baynes, 2001). For instance, Cochran contends that

> The emergence of international public spheres is in many ways contingent. However, the prospects for the incorporation of marginalized groups in world politics and the possibilities for improving beyond the actually existing institutions of international practice provide compelling reasons for pursuing them further. (Cochran, 1999: 272)

Her survey of the discipline reveals a number of topics that are generally undertheorized, including the global distortion of democratic participatory parity, the roles of non-state actors as sources of international ethical critique, and the possible construction of alternative institutional forms to the state (270). The relevance for public sphere theory is evident, leading Cochran to recommend that future research should "give some indication of how publics become institutionalized. It must inquire as to how public spheres could exercise power in international politics and when they should seek to exercise that power" (271).

These are the precise concerns of this inquiry. It is necessary to formulate a concept of extraterritorial publicity from which to proceed. In this regard, a critical review of the extant literature is instructive. Transborder deliberation is variously referred to as a "virtual public," a "global sphere," or "transnational publicity"; nevertheless, the points of divergence amongst extraterritorial public sphere theorists are more substantial than mere jargon. They can be classified into several thematic categories, involving differences over the definition of extraterritorial publics, the multiple public spheres model, the role of governance frameworks, the relevance of global inequalities, and the normative emphasis of public sphere inquiry. Each of these themes is examined respectively, before I focus on the contribution of James Bohman to aid in the construction of an alternative approach.

3.1.1 Definition

In an Introduction to an edited collection, Guidry et al. sketch out a Habermasian-influenced approach to extraterritorial dialogue. They submit the following definition of the "transnational public sphere": "a space where residents of distinct places (states or localities) and members of transnational entities (organizations or firms) elaborate discourses and practices whose consumption moves beyond national boundaries" (Guidry et al., 2000: 7–8). This rendition is inadequate because it contains no reference to basic deliberative norms that are indispensable to critical publicity. In addition, it is debatable whether corporate actors are compatible with the public sphere's guiding principle of citizen empowerment.

Nevertheless, Guidry et al. make some useful observations about the interplay between the local and the global in extraterritorial dialogue. Co-opting Giddens' famous phrase, they describe the transnational public sphere as a form of "action at a distance," and argue that its operation is frequently manifest through the connexions between local, national, and international politics. In an increasingly interconnected world, local social movements have the potential to become a global cause célèbre. Likewise, global events and discourses can often evince local political responses. Guidry et al. muse that "[t]he interactions between transnational and local public spheres, movement actors, their antagonists, and different actors within the state can be quite complex…challenging our understanding of the roles movement actors might play in politics, the sovereignty of national states in light of international conventions, and even the nature of the boundary between the state and the public sphere" (9).

However, they caution that their conception of the transnational public sphere "looks suspiciously Western," universalist, and biased in favor of certain social movements (ibid). They propose a number of caveats to temper these impressions. This includes the apprehension that globalization is an historical contingency rather than a process driven by a specific teleology. Far from inevitable, globalization is indeterminate and has no fixed future direction. As such, the transnational public sphere is "characterised by a measure of contest and contingency that is difficult to recognize, or at least emphasize, under a rubric of rationalization" (ibid.). Rather it is a volatile arena that may spill over into violence as well as lead to more progressive outcomes. Moreover, globalization is not a homogenous process, but instead is experienced by differing peoples in heterogeneous ways (and thus better understood as "globalizations"). Guidry et al. further advise that "one should attend to the variety of ways in which social movements enter the transnational public space, are potentially transformed by the encounter, and perhaps even influence globalizations themselves," and be cognizant of "the transformation of normative issues and identities in various globalizing discourses" (12–13). In other words, the global and the local are mutually transformative through the transnational public sphere. They suggest that an appropriate entry point for further investigation would be case studies of transnational social movements located in this nexus.

Guidry et al. provide some logical recommendations for future inquiry, notwithstanding the definitional inadequacies regarding the public sphere. Their approach is underdeveloped, but reflexive, and their account of "globalizations" is historically informed and sensitive to structural inequalities. They allude to the relationships between methods of communication, transnational civil society and international institutions. Speculation about the relative influence of each of these categories is certainly best grounded in case study evidence, which could also impart insights into the emancipatory potential of publicity.

3.1.2 Multiple Publics

Guidry et al. use the term "transnational public sphere," which suggests an all-encompassing arena of debate that sits uncomfortably with postmodern sensitivities about the plurality of identities and discourses in society. But Guidry et al. recognize this variegation. However, they insist that

> we need to retain that very tension between the proliferation of diversity, through multiple publics, and the homogenization and globalization suggested by a single transnational public sphere in order to recognize the ways that social movements can generate contingencies, transformations, and reconfigurations of both identities and power. (11)

Here, the argument seems paradoxical and does not convince. The hybrid model they propose is inspired by Calhoun, who conceives of multiple publics in the nation-state that are subsumed into a wider sphere defined by political borders (Calhoun, 1995). As Guidry et al. acknowledge, "[t]he coherence of a common public sphere that invites multiple publics to participate is…a difficult concept to grasp," and indeed, "[w]ithout the unity afforded by the nation it appears nearly impossible" (Guidry et al., 2000: 11). I would suggest that this requirement is too ambitious for an exploratory investigation of extra-territorial deliberative spaces. A multiple spheres model would better accommodate the heterogeneity that Guidry et al. describe, in a way consistent with propositions elsewhere in their argument that stress diversity (for example, the notion of "globalizations"). Further, it could do so without the associated conceptual difficulties of the singular public sphere model, which can have the effect of marginalizing particular issues and oppressing certain sections of society. The multiple spheres model does not preclude dialogue between publics. Indeed, regardless of its provenance, if public discourse is normatively structured, it should be addressed to an indefinite audience. Publics should not become cliques, disconnected, and shut off from the rest of the world. This would run counter to the norms of publicity, which Guidry et al. neglect to enshrine in their definition of the "transnational public sphere."

The drawbacks of the singular public sphere model are further illustrated with reference to the recent work of Colin Sparks. Contra to Guidry et al., Sparks is skeptical of the notion of the "global public sphere." He has framed his investigation of the validity of the global public around the following hypotheses:

> **P1** We would expect to find that the media that constitute the global public sphere display at least as much universality in terms of availability and access as do the existing media of state-limited public spheres…
>
> **P2** We would expect to find clear evidence that built into the media of the global public sphere are mechanisms which tend to lead it beyond any current limitations it may display. (Sparks, 1998: 112–113)

In terms of traditional mass media, Sparks argues that a "global public sphere" does not exist because, notwithstanding the common misconceptions about the extent of media globalization, most outlets remain markedly state-focused (also see Sparks, 2001: 89). Sparks acknowledges that there are certain exceptions such as the BBC World Service, but maintains that such international audiences are miniscule and bear little comparison with the mass audiences of state-based media. Further, he points out that the audience of this limited "global" media is disproportionately concentrated amongst the affluent, well-educated, anglophone elite. These disparities fall far short of the ideals of universality and equality, which leads him to discount hypothesis P1. He contends that there is as yet inconclusive evidence as to whether there are any self-correcting mechanisms built into the global media that will tackle worldwide communicative inequalities, which negates hypothesis P2. Moreover, he suggests that global disparities are so firmly entrenched that it would take many years for such a mechanism to emerge and begin to take effect. The global elites are "even more sharply differentiated from the mass of the population than were the bourgeois participants in the eighteenth-century coffee houses that formed the original inspiration for this concept" (Sparks, 1998: 121).

With regard to new media, Sparks observes that optimistic projections of the future global diffusion of ICT are not likely to be realized in the context of persistent economic inequalities (Sparks, 2001). Applications such as the Internet are developing marked tendencies toward commercial domination rather than genuine democratic deliberation (ibid.) He concludes that the term "global public sphere" should be abandoned since it is "manifestly inadequate" to capture central characteristics of the world communication order (Sparks, 1998: 122). He proposes an alternative descriptive term: "The one that fits the evidence best is 'imperialist, private sphere' " (ibid.).

Sparks' critique of the political economy of communication is a welcome antidote to the neoliberal hyperbole about the positive links between technological development and social progress. However, his definition of the public sphere in the singular has problematic implications. A unitary public sphere is a critical plank of his broader thesis. He posits that "[w]e cannot use the term 'public sphere' in anywhere near its full sense to describe a situation in which there are systematic exclusions, upon whatever grounds, of whole classes of citizens" 112). However, as Fraser demonstrates, the concept of a singular public sphere is not only inaccurate, but is prejudicial toward marginalized groups (Fraser, 1992). The singular sphere model validates the dominant public and delegitimizes the deliberative practices of counterpublics. An alternative multiple spheres model is better designed to capture the diversity and plurality of discourses in a complex society. Indeed, sociological pragmatism would direct one to expect intense fragmentation of discourse at the transnational level.

3.1.3   Governance Frameworks

Sparks' outline of the conditions of possibility for the emergence of the "global public sphere" does not include reference to the overarching political-institutional framework. However, the relation between public discussion and state activity is integral to public sphere theory. This leads one to question how it is possible to theorize about a public sphere disassociated from the state. Sparks does not demonstrate that critical publicity and meaningful deliberation is attainable without being embedded within a centralized structure of political authority. The problem is compounded by the relatively restricted definition of globalization which prefaces his discussion, which privileges the "symbolic [as] the prime site" (Sparks, 1998: 110). Little analysis is included of transformations in the global architecture of governance, and associated implications for state sovereignty. Rather, Sparks' conception of an extraterritorial public is media-centric and informed by cultural analysis. There is limited guidance here for theorizing about the function of public spheres as a means of discursive engagement about processes of governance. In contrast, these issues are central to Martin Köhler's account of the transformation of national publics. He argues that a "transnational public sphere" cannot be identified, because "the very concept of the public sphere is intrinsically bound up in structures of authority and accountability which do not exist in the transnational realm (Köhler, 1998: 233). Köhler considers how the complex social, cultural, economic, and political facets of globalization enmesh societies in new political relationships and challenge the territoriality of national political apparatuses (234). Globalization is a highly irregular process and uneven in its effects, which means that strategies of state adaptation for coping with this new environment will vary widely. Nonetheless, Köhler argues that there is an increasing tendency for state policies and institutions to pursue intergovernmental cooperation and to be responsive to transnational civil society. With respect to the latter, he maintains that most civil society remains organized around the nation-state despite the increasing visibility of international coalitions of activists. The institutional coherence of the state has not yet been superseded by similarly robust structures, but Köhler is prepared to countenance the possibility that future foundations may be laid "for an integrated global public sphere in which the distinctions between state and non-state actors may eventually be overridden" (247). "Yet as long as the state continues to be the only site of political authority in international relations," he argues, "it is impossible for a transnational public sphere—in which a global politics would have to be embedded—to emerge" (233). Instead, it is proposed that globalization is encouraging the development of a "cosmopolitan public sphere," that "reflects the ongoing centrality of the state for civil society actors, and acknowledges that state politics is the result of transnational coalition-building and interest aggregation" (ibid.). Through institutional transformation and increasing involvement in multilateralism, the state effectively "provides the impetus for the cosmopolitan enlargement of its own public sphere" (234).

Köhler's analysis captures the symbiotic relationship between the legitimation of political authority and the formation of public opinion in a public sphere. He rightly contrasts the relatively consolidated, centralized, and coherent political apparatus of the state with the fluidity and complexity of nascent transnational structures. State sovereignty may not be as secure as he suggests, but his open-ended approach invites the reappraisal of the evolution of global governance and its effects on national autonomy. Where Köhler provides little comment is on the role of ICT, which is not treated as a variable worthy of distinct analysis. However, ICT is deeply implicated in processes of globalization, and the direction of its development has a direct bearing on the deliberative capacity of citizens. Investigation into the global diffusion and political economy of ICT would yield profound insights about the potential for extraterritorial public spheres.

The difference in emphasis between the approaches of Sparks and Köhler reflect the academic orientation of the authors (Communication Studies for the former, IR for the latter). Interestingly, the gaps in their analyses are complementary, and indicate the potential for synthesis between these disciplines. I shall revisit this idea in section 3.1.6.

3.1.4   Global Inequalities

A recurring theme in public sphere theory is social inequality. To some extent, all public sphere theorists must consider whether entrenched multifaceted inequalities (economic, social, technological, etc.) can be reconciled with the norms of publicity. More specifically, Nancy Fraser once challenged Habermas by asking whether capitalism was compatible with a "non-exclusionary and genuinely democratic public sphere" (cited in Habermas, 1992a: 468). We have seen how these concerns are paramount for Sparks, but they are accredited with less significance by other theorists. For example, Hauke Brunkhorst adapts Fraser's definition of a deliberative sphere to argue that a "weak global public" has existed since the League of Nations, and especially since the foundation of the UN (Brunkhorst, 2002: 680). Nonetheless, he does not mirror Fraser's focus on the implications of societal inequality.

Brunkhorst theorizes that: "[t]he constitutional precondition of this weak public is realized in the existence of a core of binding legal rights and general principles of international law. Its social precondition is enabled by the media of global communications and by a transnational network of associations" (ibid.). Brunkhorst envisions this forum as a potential "strong public in the making" if it is strengthened by a framework of norms in a developing global constitution. He posits that the necessary conditions of a strong public are "the existence of a working system of hard-law human rights embedded in a well-ordered global society. It's sufficient condition is a public sphere enabled technologically by electronic media in interplay with associations and individuals that make communicative use of these" (690). For Brunkhorst, public spheres are anchored on ICT and the constitutionalization of global governance.

These enabling conditions are effectively mapped out, yet I would argue that they are not the only significant variables. It is unfortunate that Brunkhorst devotes little attention to implications of global structural inequalities to communicative resources. The "weak global public" that he describes is likely to be mainly representative of the transnational elite—and so it will not be as inclusive or emancipatory as the ambitious moniker suggests. This indicates another divergence apparent in public sphere theory—theorists place differential emphasis to certain normative concerns. The problematic implications of a restrained approach to norms are best illustrated with reference to Marc Lynch and Jennifer Mitzen.

3.1.5   Normative Emphasis

Marc Lynch has delivered the most consistent and rigorous application of public sphere theory to IR to date, which has been a formative influence on this inquiry (Lynch, 1999, 2000). Drawing on the work of Habermas, Lynch "argues for placing communication at the center of International Relations theory," and claims that a public sphere approach can synthesize insights from rationalism and constructivism (Lynch, 1999: 3, see also 9–13). Lynch conceives global politics as structured by traditional forms of "strategic interaction" (resembling the market) and a public sphere of "communicative action" (resembling the forum) based on deliberation, dialogue and persuasion. With regard to the latter, Lynch is primarily interested in how the presence of international public spheres can be detected by the way state behavior is structured through the act of giving justifications (Lynch, 2000: 318). He describes a tension between deliberation and international anarchy, meaning that only nonbinding consensus can result from international public sphere dialogue (Lynch, 1999: 37). Thus: "[t]he manipulation and contestation of an international consensus takes the place of the effort to influence state policy as the defining characteristic of public activity" (47). This is a generally useful means to conceptualize the role of public spheres in the absence of an overarching sovereign body: as a location for the forming of norms and the construction of interests and identities (322). However, it also implies the absence of governance structures at an international level that can take authoritative decisions. In certain cases, the division of political responsibility amongst domestic and international agents may be rather more fluid and complex than Lynch recognizes. Moreover, if trends toward global governance continue to become more pronounced, state authority will be increasingly disaggregated to a wider range of actors. This raises interesting questions about the conceivable political role of extraterritorial public spheres in addressing the democratic deficit of contemporary governance. But speculation about immanent emancipatory possibilities is difficult without a direct focus on global political-institutional transformations.

Lynch defines a public sphere as "a site of interaction in which actors reach understandings about contentious issues of shared concern through the public exchange of discourse" (Lynch, 2000: 316). This functional definition purposely makes no reference to location or to the content of discourse, which enables one to generalize across state borders and cultural differences. It also allows for multiple, overlapping spheres of state and non-state actors. The historical specificity of the Habermasian version has not been retained, but wider applicability is necessary in the context of an infinitely differentiated audience (Chapter 2). Lynch supplements this definition with reference to fundamental norms of publicity: for example, debate should be as inclusive as possible, oriented toward consensus, and free of direct appeals to power. The normative kernel of public sphere theory has thus been preserved while other distinctive elements have been shorn out of necessity. As such, Lynch's definition has practical heuristic benefits in investigating new transnational sites of deliberation.

Lynch recommends that "[n]o prior theoretical decision should be made which arbitrarily close off participation in international public spheres; what is necessary is to determine the political implications of different participation/exclusion rules and practices" (324). I concur with this assertion; and so have normative misgivings about privileging states as public sphere actors. According to the Habermasian rendition, the public sphere is a site separate from sovereign bodies, whereby citizens can exercise democratic constraints on political authority. The revolutionary promise of public sphere theory may be compromised if citizens are not part of the picture. Lynch has explored U.S. foreign relations, the Jordanian and Arab public and the need for cross-cultural communication after 9/11; all have yielded fascinating insights (e.g., Lynch 1999, 2002, 2003, 2005). Yet the focus in his oeuvre largely remains on the nation-state. The result is that Lynch's social critique is somewhat inhibited, and the normative potential of his approach consequently restricted. Lynch himself admits that: "I have hardly engaged with the normative theory that has been the primary application of Habermas in IR theory to this point" (Lynch, 2000: 268), emphasizing instead the "pragmatic and empirical implications of deliberation" (320). Although more ambitious goals "inform the underlying agenda," there is a danger that they are effectively side-stepped (ibid.).

Following Lynch, Jennifer Mitzen reconstructs Habermasian theory to analyze deliberative processes between states (Mitzen, 2001, 2005). She draws a distinction between "weak" publics of non-state actors, and an "interstate public" of state actors. The latter embodies a "self-regulating" dynamic that she believes can help to mitigate the security dilemma. Mitzen asserts that most extant extraterritorial public sphere theory describes the activities of citizens and transnational non-state actors that seek to increase the public accountability of state power. In this literature, "legitimation is assumed to take place only among citizens and directed towards states, rather than among states themselves" (Mitzen, 2005: 402). Legitimation is a vertical process in "weak" publics, but Mitzen contends that "interstate publics" contain a horizontal legitimation dynamic, which has been undertheorized. She critiques public sphere theory for this neglect, and for bracketing order from legitimation, claiming it is not clear "how to contain the instability of communicative action where argument is not backstopped by either a shared lifeworld or positive law" (404). For Mitzen, international anarchy breeds an atmosphere of insecurity and mistrust, which is inimical to public sphere dialogue. However, she points to the Peace of Westphalia and the Congress of Vienna to exemplify how these obstacles can be surmounted, through international society and face-to-face conference diplomacy.

Mitzen's analyses are helpful in conceptualizing interstate deliberation, and a pertinent reminder that international public dialogue has a long historical pedigree. However, her emphasis on the importance of order and of centralized enforcement powers may bias her investigations against nascent publics that operate outside of established legal frameworks, or are subject to political persecution rather than legal protection. As discussed in the previous chapter, such counterpublics have historically proven to be potent sources of societal transformation. The state-centricity of Mitzen's approach also has adverse normative implications. State boundaries promote categories of inclusion and exclusion that can be antithetical to human emancipation. Interstate dialogue may not directly relate to the emancipatory interest of citizens. The critical force of public sphere theory rests on the empowerment of public reason vis-à-vis political authority. If this connection is weakened or lost, the radical potential of public sphere theory is dramatically curtailed, and its progressive purpose diluted. This is not to dismiss attempts to seek greater understanding of the processes surrounding interstate consensus, but rather to restate a personal commitment to a more cosmopolitan approach.

Many of these issues have been satisfactorily resolved by James Bohman, who has authored some of the most thoughtful and theoretically informed reflections on extraterritorial publicity. Insights gleaned from a critical appraisal of his work suggest the next steps forward for public sphere theory.

3.1.6 Toward a Synthesized Approach

Unlike Sparks, James Bohman is prepared to countenance the possibility of an immanent global public sphere—he considers the notion unlikely, but not unattainable (Bohman, 1998: 201). In order to theorize about these issues, he argues that the traditional Eurocentric notion of the public sphere needs to be challenged, and an alternative cosmopolitan definition adopted, "that takes into account manifestly different historical conditions" (201). He maintains that the bourgeois or domestic public sphere is not the only possible manifestation of critical publicity; there are many different empirical variants. Bohman propounds an exploratory approach with a definition of a public sphere sufficiently broad to be historically generalizable and sensitive to different cultural contexts (201–202). This suggests a functional definition similar to Lynch's, but Bohman's emphasis on the norms of publicity is much stronger. Normative concerns are preeminent rather than merely tangential to analysis. Bohman insists that "public spheres must still be locations for social and cultural criticism" (202). He asserts that higher levels of publicity are realized when communication is directed at an indefinite audience and the roles of speaker and listener are exchanged in an egalitarian manner. All participants should be prepared to be responsive and answerable to the concerns of others (207–208).

On the basis of this understanding, Bohman counters unsubstantiated assumptions about the "global public" by meditating on the institutional prerequisites for such a sphere. The conditions that he identifies are closely aligned with Habermasian theory and provide a clear alternative theoretical framework

> For a global public sphere to be possible, three conditions must be met: the existence of a mass media of global scope and equipped with the technological capacities of speed of communication; the emergence of

a variety of transnational and local public spheres and sub-publics which organize their own audiences and develop their own forms of publicity; and finally, the requisite organization and institutions of civil society, the state and international organizations which support and make possible a variety of public spheres. (201)

Bohman argues that these conditions bear poor comparison with present reality. The orientation of global media and the embryonic state of transnational civil society and international political institutions militate against the realization of a global public sphere. He also points out that deliberation in such a vast social space would be hindered by the lack of cultural specificity: "the larger an unstructured and undifferentiated audience is, the less likely it is that to become part of it requires the public use of reason" (211). Bohman sees communications in this context as unlikely locations for social critique, since the audience are more likely to be "anonymous" to each other and to the producers of media content. He argues that addressees of such anonymous communications constitute an indefinite audience "in a purely *aggregative* sense: it is not an idealized audience that it addressed, but the aggregate audience of all those who can potentially gain access to the material and interpret it as they wish" (ibid., original emphasis).

Elsewhere, Bohman extends this problematique to ICT. He perceives anonymity as intrinsic to Internet communication: "In a network mediated by a computer interface, we do not know who is actually speaking; we also do not know whom we expect to respond, if they will respond or if the response will be sustained" (Bohman, 2004: 138). Paradoxically, the Internet may offer increased opportunities for interactivity, yet diminish chances of responsive uptake (135) The difficulties of anonymity in extraterritorial publics are thus exacerbated. Identities tend to be explicit in face-to-face or print publics; Bohman maintains the normative expectation of electronic communication is different when the authenticity of statements cannot be ascertained.

The Internet can be utilized in varying ways and Bohman recognizes that software can alter the conditions of dialogue: for example, intranets and firewalls may be used by corporations to effectively privatize public space (140) This indicates the integral role of institutions in defining the public character of the Internet. Certainly the direction of its future development is far from fixed and struggles over contesting interpretations of publicity are ongoing between corporations and civil society (141). Bohman describes the former as powerful intermediaries, and calls for organizations in civil society "that have becomes concerned with the publicity of electronic space... [to] seek to create, institutionalize, expand and protect it" (143). Emancipatory potentialities are located within the counter-intermediary role of civil society, which directs our political attention to the mediated communications of transnational social movements. Bohman rightly rejects technological determinism, instead stressing the role of social agents in shaping the Internet's contribution to civic life.

He explores the potential function of agency by identifying the gradual transformation of local public spheres through transnational civil society into cosmopolitan forms of publicity, where two or more delimited public spheres interact (Bohman, 1998: 213). These spheres have the potential to broaden as participants develop the public use of reason to negotiate across boundaries and different cultural identities. In this sense, cosmopolitan spheres can be considered global in the sense that members are committed to the norms of publicity requiring that "they hold themselves accountable and answerable to all who use their reason publicly, and thus to all citizens of other public spheres" (214). Because the inherent features of the medium compound the difficulties of transnational dialogue, publics are more likely to expand through the activities of transnational civil society (Bohman, 2004: 138). Bohman clearly expresses the thorny implications of the anonymity of Internet communication, but the process whereby cosmopolitan spheres progress to higher levels of publicity is unclear. How can we understand the motivations of a group of geographically disparate citizens to engage in normatively structured discourse? There seems to be a need for another analytical category to describe the relevant social factors that facilitate the self-identification of a public.

This becomes apparent if we return again to consider the features of a state-based public. Participants were united by citizenship status. All lived within the same territorial boundaries and were subject to the same sovereign authority. This common experience of governance provided a clear justification for the norms of publicity: the democratic belief that the legitimacy of sovereign authority derives from deliberation that is inclusive of all affected actors. An extraterritorial public sphere is not characterized by a similar correspondence of social and political boundaries. But public sphere discourse will not be possible beyond the state unless actors have a basis from which to engage with one another in normatively structured dialogue. This will require a minimal sense of commonality amongst the interlocutors; some kind of "...sense that they belong to the same social and political entities, despite their differences" (Dahlgren, 2005: 158).

Dahlgren describes as a feeling of "affinity," grounded on a realization amongst different groups that they "have to deal with one another to make their common entities work, whether at the level of the neighbourhood, nation state, or the global arena" (ibid.). I find this notion a useful one to supplement Bohman's conception.[1] It also may serve to mitigate his pessimism about online deliberation, or in divergent cultural contexts (assertions that he does not support with substantive evidence). The notion of mutual affinity also suggests that similar democratic norms can underpin national and extraterritorial deliberation, and implies the pivotal role of governance frameworks in framing public debate (although Dahlgren does not explore this in great detail).

In an earlier work, Bohman made some interesting observations about the relationship between deliberation and governance institutions when he elaborated on the role of "cosmopolitan public spheres." He argued that the public must not only perform a "scrutinizing" function, but also have "agenda-setting" influence if dialogue is to be meaningful (Bohman, 1997: 183). It is a valuable observation. Public opinion should reflect the common interest of those affected by political authority, and decision-makers should be responsive to these concerns (187). Otherwise, public deliberation is little more than ineffectual talk. Some kind of relationship with governance is essential if critical publicity is to have political force.

The "agenda-setting capacity" of the public also demands that citizens must "be dynamic enough to reshape the framework of existing political institutions to require acknowledgement of the rights of members of the universal community outside the boundaries of its territories and membership" (ibid.). Hence there is a logic toward greater inclusion. This reasoning chimes with the analysis of Guidry et al. regarding the mutually transformative relationship between the local and the global through public spheres. Bohman points to the civil rights movement as an example of where the public thinks and acts self-referentially, and transforms the conditions of political deliberation as it undergoes a process of self-transformation (192). He asserts that the public is able to change political institutions indirectly through dialogic engagement with sites of political authority. "In the process," Bohman argues, "institutions are changed in a variety of ways: in their concerns, in their ongoing interpretation of rules and procedures, in their dominant problem-solving strategies, and so on" (191). Such transformation is necessary if institutions are to retain legitimacy.

Bohman presents an elegant delineation of the prerequisites of extraterritorial public spheres. His is a theoretical account, with little empirical evidence, but it offers substantially robust grounds for more detailed investigation.

3.1.7 Precepts for Public Sphere Research

The above review has highlighted a number of common limitations, shortcomings and ambiguities in the extraterritorial public spheres literature. There are also instances where concepts are usefully framed and theory well-designed. Criteria can be derived from this appraisal to aid the construction of a synthesized critical-theoretical approach.

First, the unsubstantiated claims of Dryzek et al. should be avoided: it cannot be assumed that extraterritorial public spheres are extant. Instead, a conceptual framework should be formulated as a heuristic aid for research into present conditions. As Lynch argues, a functional, generalizable model of extraterritorial dialogue is appropriate for exploring the possible emergence of public spheres. This permits sensitivity to different cultural contexts, as the common characteristics of discourse are likely to decrease in specificity as the public become increasingly diverse.

Second, Sparks demonstrates that it is important to be attentive to the impact of global disparities, as issues of inclusion and exclusion are central to normative public sphere theory. Brunkhorst's otherwise excellent inquiry did not give due consideration to global inequalities and was underinformed as a result.

Third, the singular sphere model of Guidry et al. and Sparks is inappropriate at the fragmented and heterogeneous transnational level. A multiple spheres model is more realistic. As it is better able to accommodate plurality, it also has the advantage of being less exclusionary. The term *"transnational public spheres"* perhaps best reflects this non-state-centric, multiple spheres approach.

Fourth, the way in which membership of public spheres is conceived has a bearing on the radical potential of theory. Lynch has tended to study state-interaction, as does Mitzen. Guidry et al. conceive of corporate participants in the public sphere. However, Bohman's focus on non-state actors and citizen empowerment bears a more faithful correspondence to Habermas' normative concerns. Bohman also delineates three conditions of extraterritorial public spheres—media, publics, and governance institutions—which broaden the analytical frame to new political and social processes beyond the nation-state. These conditions map roughly onto Habermasian public sphere theory.

Fifth, the motivating factors and interrelationships between potential participants in emergent transnational public spheres should be explored further, and Dahlgren's notion of "affinity" is promising in this regard. Lastly, if transnational public spheres are politically significant, we should expect to discern their transformative influence on the institutional environment. Bohman describes this in terms of influencing the political agenda; Lynch conceives the public's role in terms of manipulating and contesting international consensus, and the construction of interests and identities. Guidry et al. suggest that the local and global can be mutually transformative, and that these effects are best perceived through case study evidence. This is eminently sensible means to pursue further inquiry; for instance, Bohman's speculations could be usefully supplemented with empirical evidence. I want to propose a method whereby all of these concerns can be integrated.

3.2 Preconditions for the Emergence of Transnational Public Spheres

Since *STPS* was published, there have been extraordinary developments in communication technologies that are associated with transformations in world order. The question is what effect, if any, these developments have had on public spheres. Habermas can be criticized for being too conservative in his speculations about the potential of the public sphere. The enormous intensification of cross-border communicative interaction, which creates shared spaces of experience and argumentation, has made the transnational dimension almost impossible to ignore when theorizing about public spheres. Such theoretical activity does not necessarily rely upon any a priori claim to actually existing transnational public spheres. But this inquiry *does* assert the potential for extraterritorial deliberative spaces to engender rational-critical dialogue orientated toward consensus.

The concept of transnational public spheres is related but not directly analogous to local or national public spheres. Therefore, following Lynch and Bohman, I propose a functional definition of a public sphere, which makes no assumptions regarding location or the substantive content of dialogue.

A transnational public sphere can be understood as a site of deliberation in which non-state actors reach understandings about issues of common concern according to the norms of publicity.

Note that this definition allows for the existence of multiple, overlapping spheres. Contra Sparks, I contend that the existence of counterpublics in the early-modern era suggests similar, if not greater fragmentation in the contemporary transnational realm. Of prime importance are the normative requirements of rational-critical debate orientated toward consensus, such as the inclusion of all affected actors in deliberation, the nominal equality of all participants, and the adjudication of competing claims through reason rather than recourse to power. These criteria can be summarized as the norms of publicity. Although this definition may not have the advantage of Habermas' historical specificity, Bohman has demonstrated the need for generalizability at the transnational level, in order to incorporate the diverse experiences of non-Western cultures. The norms of publicity also imply that social disparities are key concerns, since they will affect the means of access and participation in a public sphere.

In the previous chapter, I identified three basic conditions of possibility for public spheres from a reading of *STPS*. These were as follows: ability to communicate (via the medium of print), separation from public authority (i.e., the state serving as the addressee of public sphere dialogue), and adherence to the norms of publicity (which requires a sufficient degree of affinity between participants to engage in normatively structured discourse). Extrapolating from this, I propose that three similar structural preconditions could provide for the emergence of transnational public spheres: transborder communicative capacity (via new media), transformations in sites of political authority (varied global governance structures acting as the addressee/s of public sphere dialogue), and transnational communities of mutual affinity (as with the domestic counterpart above, only the basis for mutual affinity would rest on a foundation other than shared territory or national citizenship). These trends can be understood as interrelated processes of globalization. A transnational public sphere could manifest if there were a confluence of all preconditions around a certain issue-area. Let me sketch each criterion respectively.

First, transborder communicative capacity refers to all media with transnational reach, but particularly new ICT such as the Internet. Digital and networked technologies are qualitatively different from their analogue and mass media antecedents. They are interactive and decentralized, with increasing capacity for user-generated content. They are an extraordinarily fast and efficient means of information dissemination, and they are proliferating at an unprecedented extent. ICT represents enhanced opportunities for grassroots dialogue and political mobilization across state borders. However, these prospects

are threatened by gross inequalities in access and ownership of ICT; and restrictions on freedom of speech by repressive governments and a global media oligarchy.

The second precondition is transformations in sites of political authority. Although states still retain the supreme legal claim over the exercise of authority within their own borders, this has been complicated in recent years by a number of factors, including the development of international law and the complex identities of citizens. The term "global governance" describes the spectrum of distributed responsibility amongst actors and regimes that form the rules and norms of world order. The increasing prominence of such governance mechanisms calls into question orthodox presumptions about the sovereignty and authority of the nation-state. These transformations may indicate emerging institutional preconditions for transnational public spheres, albeit at an embryonic stage.

The dynamic relationship between each of the above factors can be examined by studying transnational social movements as representative of potential sites of struggle in late modernity. Therefore, the third condition, transnational networks of mutual affinity, refers to groups of citizens that mainly communicate via ICT, especially to engage in political activism. There are indications that traditional forms of national politics are related to patterns of civic disengagement, as evidenced in many established democracies by a pattern of decreasing electoral turnout and party membership. At the same time, we are witnessing a rise in a new political culture of citizens directly claiming interests and entitlements, sometimes at the transnational level, through single-issue pressure groups. The participants in these social movements may attempt to achieve their aims at the local, national, or transnational level, defining meaningful political borders as they see fit. As Köhler observes: "The borders of the political community become meaningless in the traditional political sense of mobilization for specific goals. The new borders are differences of language and political culture, and they do not necessarily combine with community borders" (Köhler, 1998: 238–239). Through these contestatory processes, we may perceive evidence of transnational communities bound by sentiments of mutual affinity.

As Dahlgren argues, a public sphere rests on feelings of "affinity" amongst its members, meaning that there should be a mutual recognition of the moral-political validity of inclusive discourse (Dahlgren, 2002: 17). Dialogue in these circumstances will be characterized by the normative conditions of publicity (e.g., inclusivity, intelligibility, accountability, reflexivity). Bohman's work suggests that feelings of mutual affinity may be frustrated by factors such as the geographic diffusion of the members and the characteristics of computer-mediated communication. Therefore, these criteria will be judged with reference to case study evidence.

The chosen case studies are drawn from three subject areas: the international women's movement, the Zapatistas, and Greenpeace. Each social movement has a different basis from which participants can derive a sense of mutual affinity, and will be treated as possible examples of emergent transnational public spheres. Bohman maintains that effectual public spheres are sites of societal transformation. I will therefore assess to what extent these social movements have influenced the mainstream political agenda and affected the international institutional framework.

3.3   Conclusion

Extant public sphere theory needs to be expanded and modified to assess the possible emergence of transnational public spheres. This inquiry attempts to systematically analyze the conditions of possibility for expanded publics. It represents a continuation of the normative and political concerns that motivated Habermasian public sphere theory. This chapter has outlined a conceptual framework, grounded in the international critical theory tradition, to aid theorization about the sociopolitical implications of ICT. Three institutional prerequisites of transnational public spheres have been proposed—transborder communicative capacity, transformations in sites of political authority, and transnational networks of mutual affinity. The copresence of each condition is necessary, so emergent transnational public spheres are most likely to be located around specific issue-areas. The chapters that follow evaluate the prospects for these structural preconditions to materialize, and for transnational publicity to be gradually instantiated.

CHAPTER 4

The Information Age: Transborder Communicative Capacity

Conventional public sphere theory is ill-placed to evaluate the import of cross-border communicative flows, as it presupposes an alliance between political territory and the circulation of dialogue. This fit once seemed so close that some have made the extrapolation that public spheres require a physical locale and proximate interlocutors. This misapprehension is perhaps partly encouraged by the terminology of public sphere theory. It is unfortunate that the imagery of face-to-face interaction is encouraged by repeated allusions to reflexive dialogue. In fact, virtuality has been a central feature of the public sphere in most of its historical manifestations, that is to say that discourse has been conducted at a distance (Warner, 2002). Mediated dialogue is a necessary feature of any large-scale, complex social organization, providing the only means of interaction between spatially dispersed actors. Therefore there is no a priori reason why transnational mediated communication should be incompatible with critical publicity.

Transnational communication dates back thousands of years, but before the mid-nineteenth century, the few transnational communication channels that did exist were the province of the aristocratic or military elites. Since the invention of the telegraph, the transmission capability and rates of global access to ICT has risen exponentially. The resulting transformation in the world's media landscape has been awesome. Our society is suffused by a dense network of information exchange and communication flows, mediated through technologies such as radios, televisions, fixed telephones, mobile telephones, personal computers (PCs), and the Internet. New media[1] such as the latter have facilitated an unprecedented explosion in the scope and intensity of cross-border communicative activity. This expansion in physical infrastructure could provide the material capacity for transnational public spheres to materialize around certain issue-areas.

A transnational public sphere rests on the ability of interlocutors to communicate across state borders with ease. It could be said that this requirement has already been met in terms of material capability. ICT has eradicated temporal and spatial barriers to distanced communication. However, the prerequisites of public sphere debate are more demanding than this; the category of "transnational communicative capacity" also entails qualitative requirements. Critical publicity must be as inclusive as possible—and so there should be wide diffusion of communication technology and maximal opportunities for access and participation. In addition, dialogue should be free and open, unhindered by censorship and undistorted by manipulative publicity from governments and corporations. Evidently, this ideal is far from met in the present world communication order. There are entrenched social exclusions and ownership of global media is concentrated to the point of oligarchy. Many governments also restrict freedom of expression and censor media content.

This chapter considers whether these factors effectively preclude the realization of transnational public spheres. It is structured into three main sections. First, there is a survey of the emergence of the new media environment, where the rise of global media corporations and the commercialization of media content are critiqued. The second section offers an analysis of corporate and state involvement in Internet censorship and citizen surveillance. Lastly, there is an outline of the multiple disparities and inequalities that characterize our supposed "information age." I conclude by acknowledging that the structural precondition of communicative capacity is only present for privileged sections of world society.

4.1 THE "INFORMATION AGE"

The advent of the Internet prompted many to herald the coming of an "information age." It is a useful term that also captures the sense of the immense penetration and ubiquity of ICTs in a media-saturated world. The impressive-sounding moniker suggests an epochal shift comparable in socioeconomic importance to past eras such as the Industrial Revolution. It also mistakenly suggests that the shift has been sudden. In fact, the new media environment is the result of an incremental transformation that cannot be ascribed to a singular innovation or one particular type of technology. It is the culmination of a long historical process that can be traced back to the invention of the telegraph. As Deibert observes, the information age

> reflects a complex melding and converging of distinct technologies into a single integrated *web* of digital-electronic-telecommunications—a process that has roots reaching back to the late nineteenth century, and that encompasses a series of technological innovations that continued through the twentieth century, culminating in the digital convergence that began in the late 1960s. (Deibert, 1997: 114, original emphasis)

Digitization has produced an intricate global network of communication infrastructure, characterized by the enmeshment of technologies that were once part of separate platforms (audio, visual, or textual). Technologies such as mobile phones—where it is possible not only to telephone others, but also to send text messages, download films, and listen to MP3 files—illustrate how different functions have been successfully integrated in one facility.

The capacity to translate a range of information into digital format and to process it through the same channels is eroding the boundaries that once used to exist between traditional media industries. Formerly, media corporations grew around discrete sectors (such as newspapers, cinema, or radio) and concentrated on manufacturing product appropriate for a specific means of delivery (news articles, motion pictures, disc-jockey shows). But such divisions are of less significance with the dawn of a "universal media." Digital convergence merges the traditional functions of computers, telephony, televisions, and other media. As a result of either initiative or necessity, media corporations are broadening their product range and diversifying their investment portfolios (Croteau and Hynes, 2005). In recent years, there has been a flurry of mergers and acquisitions, as companies in the midst of technological upheaval have sought to protect themselves from an uncertain future by investing in an ever-expanding profile. Convergence has increased the importance of the media sector to the global economy. The ICT industry has been growing significantly faster than the wider economy, with international communication growing fastest overall. It is difficult to overestimate the importance of this sector for the health of global capitalism.

It remains to be seen how such massive conglomerates will perform in the long term. But the unique position of these corporations clearly affords them a substantial structural advantage in manipulating publicity to promote self-serving commercial values. The Internet is not immune to these trends, having become irrevocably commercialized since its inception (Simpson, 2004). Companies are hugely significant providers of online content, which has tilted the online balance further in favor of mass distribution, advertising, and e-commerce (Salter, 2004: 196–197). The specific issues surrounding the Internet will be examined shortly. Beforehand, it is worth placing the Internet into a wider political-economic context by considering the global corporate structure of the industry.

4.1.1   The Rise of Global Media

Although media systems still remain primarily national or local, digital convergence and the worldwide trend toward telecommunications liberalization have bolstered the position of the global media conglomerates. The changes in global media in the last couple of decades are perhaps most apparent in the international rise of commercial television. The worldwide trend toward deregulation and privatization has triggered an explosion in global commercial broadcasting owing to the liberalization of national television systems. The immense growth of commercial television has intensified since the 1990s, at the expense of public broadcasting services (PBS). The aggressive competition poses a long-term threat to the survival of PBS in all regions of the world. The BBC's adaptation strategy to this changed environment is perhaps one of the most ambitious. The BBC is attempting to pursue global commercial activities while sustaining their public service remit at home. In recent years they have capitalized on their famous brand name by launching BBC World Service Television, the BBC Web site, and expanding BBC World Service on radio. A major project has been the expansion of existing analogue channels to digital interactive services. The commercial branches of the corporation are seen as ensuring the future survival of the BBC by providing an important source of subsidy for public service programming, whilst retaining the prestige associated with the brand (BBC, 2006). How successful the BBC will prove in this venture is still too early to tell. However, it is the type of strategy that is only open to well-funded Northern PBS—much of the rest of the world's public sector are facing a future of increasing marginalization, or future commercialization. The domestic push toward privatization and underfunding of PBS represents core themes behind the rise of global media. It illustrates how convergence and neoliberalism have helped to serve the commercial interests of the global media corporations, and how media diversity has been eroded as a result.

The United States exerts a domineering hold on the global enter-tainment industry, with no comparable export rivals in terms of televi-sion, film, and music. Some scholars warn against overstating U.S. hegemony (e.g., Compaine, 2002, 2005). Certainly, some of the larg-est U.S. firms have significant foreign ownership, and a number of the world's greatest conglomerates originate from outside the United States, such as Japanese Sony, French Vivendi Universal, and the Canadian Thompson Corporation. There are other important film industries outside of America—such as India's "Bollywood" that exceeds U.S. output, and is a major Asian supplier. Moreover, the U.S. industry has had difficult times of late: for example, there has recently been a drop in music sales owing to the rise in CD piracy (RIAA, 2006). The black market in the latter is said to have an annual global turnover of $4.5 billion (IFPI, 2006). Likewise, piracy is estimated to cost the worldwide motion picture industry $18.2 billion during 2005, with the U.S. industry accounting for $6.1 billion of the loss (MPAA, 2006). Common suppositions regarding the "cultural hegemony" of U.S. media can also be challenged. There is a notable trend toward the regionalization and localization of media content to suit the cultural priorities of audiences. Robertson calls this phenomenon "glocaliza-tion": a term that describes how Northern media adapt using new media to appeal to local languages, styles, and cultural conventions (Robertson, 1992). Chevaldonne describes this process as a

> subcontracting of market niches to local companies better equipped to deal with audiences which possess special characteristics which create special expectations at the level of message: language, the place of music and dance, history, religion, and a certain way of coding the rela-tions between the sexes, generations and social classes. (Chevaldonne, 1987: 145)

This global-local interaction can therefore be "good for business," and serve to reinforce ethnic cultural identity, rather than U.S. cul-tural hegemony.

Despite these caveats, it cannot be denied that the overall eco-nomic and cultural predominance of U.S. media persists. The six largest media and entertainment corporations in the world today are all regarded as American: General Electric, Microsoft, Time Warner, Comcast, News Corporation, and Disney (Financial Times, 2006). No other country can match this concentration of economic might and global reach. Robert McChesney refers to these companies as the "first tier" of the media industry, which are followed by around 50 or so "second tier" companies that operate on a national or regional level (McChesney, 2001). Nationally, the concentration can be even more intense: for example, over the past 20 years the number of corporations dominating U.S. media companies has contracted from 50 to just 5 giants (Bagdikian, 2004). Recent trends suggest that corporations will persist in attempting even larger mergers. For instance, Viacom attempted to buy out CBS in 1999, Comcast bid for Disney early in 2004, and AOL made the largest merger in media history with Time Warner in 2000. The latter deal represented $350 billion—which was more than 1,000 times larger than the biggest deal of 20 years earlier (ibid.). Some of the disquiet that this move precipitated was voiced by Tom Rosenstiel of the Project for Excellence in Journalism, who warned that "what this merger invites is the possibility of a new era in American communication that sees the end of an independent press" (BBC, 2000).

For Herman and McChesney, these firms are no less than the "new missionaries" of corporate capitalism. They conceive of their influence thus:

> As the media are commercialized and centralized, their self-protective power within each country increases from the growing command over information flows, political influence, and ability to set the media-political agenda (which comports well with that of advertisers and the corporate community at large). (Herman and McChesney, 1997: 9)

These themes resonate strongly with the thesis of public sphere degradation in *Structural Transformation*. Habermas demonstrated the corrosive effect of the overweening influence of large media corporations on critical publicity. Deliberation was progressively distorted and manipulated to serve commercial interests. If not adequately counterbalanced, the continuing conglomeration of global media is likely to perpetuate this process.

Media concentration can mean that companies can act together as an oligopoly or cartel. They will have a common interest in avoiding public scrutiny of their actions. It is a democratic necessity to ensure such powerful actors are publicly accountable, and of heightened importance considering the major defense interests of companies like General Electric. Cartels also work together to marginalize their competitors to consolidate their stranglehold on the market. They are likely to have an ideological interest in filtering out counterhegemonic discourse that they find threatening or unpalatable. Sometimes the owner may exercise this power by covertly or overtly compromising editorial independence; for instance, it is commonly thought that undue prominence is accorded to the political opinions of Rupert Murdoch by his newspaper stable in the United Kingdom. But often, distortion of publicity arises through the structural effects and commercial pressures of the media market, which conditions editors and journalists to prioritize certain issues and to neglect others. Naturally, sensationalist stories that attract prurient interest are likely to be high profile, and stories that could damage the profitability of important advertisers will tend to be shelved. Counterhegemonic ideas are also unlikely to be treated seriously because they do not easily complement the values of hyper-commercialism embodied by the media cartel. Miller considers the consequences of this trend for the United States:

> Of all the [media] cartel's dangerous consequences for American society and culture, the worst is the corrosive influence on journalism. Under AOL Time Warner, GE, Viacom et al., the news is, with a few exceptions, yet another version of entertainment that the cartel also vends nonstop. This is also nothing new—consider the newsreels of yesteryear—but the gigantic scale and thoroughness of the corporate concentration has made a world of difference...the news divisions of the media cartel appear to work *against* the public interest—and *for* their parent companies. (Miller, 2002: 13, original emphasis)

Moreover, global media moguls have such a potential degree of influence over media content and distribution that they can claim to have a sizeable role in influencing public opinion, and therefore have disproportional leverage with governments. The concern of free speech advocates is that greater media concentration translates into less diversity of expression, fewer dissenting voices, and thus fewer opportunities for meaningful debate. The output of media conglomerates reflects such a concentrated ownership base that it is unlikely to fairly reflect the diverse range of society's needs, values, opinions, and ideas (particularly of marginalized and subordinate groups). It restricts the space and opportunity for governing orthodoxies to be exposed to challenge. News coverage may be particularly skewed and political bias may be apparent in reporting, which can have the effect of distorting public debate. These concerns appeared to be evidenced by the studies conducted by Fairness and Accuracy in Reporting (FAIR). For example, in the run-up to the 2003 Iraq War, the major networks only gave 3 percent of on-camera news coverage to U.S. sources representing an antiwar stance. For CBS, the figure was less than 1 percent. Yet this was at a time when opinion polls were consistently registering that around 27 percent of the American public were opposed to the war (Rendall and Broughel, 2003). Across the corporate media, there was a chronic lack of critical analysis of the government's claims before, during, and after the invasion; and substantial underrepresentation of voices that differed from the official agenda (FAIR, 2007).

Distortion of public debate by powerful corporations runs counter to the ideal of a public sphere representing an open realm for discussion, free from manipulation by partisan economic forces. This as an ideal may be ultimately unobtainable, but it provides a normative paradigm against which actual circumstances can be measured. Current conditions fall far short of this ideal. The predominance of neoliberal, profit-motivated, corporate interests behind the development of global media signals the further degradation of publicity. However, this pessimistic portrayal of public sphere decline must be balanced against encouraging signs that the expansion of global media has dramatically expanded transborder communicative capacity for millions of people—even those in the poorest countries of the world. With specific reference to the Internet, the remainder of this chapter evaluates whether an expansion of communicative capacity can be said to simplistically equate to an expansion of the public sphere.

4.2 The Internet Revolution

The most iconic technology of the information age is the Internet. It is unprecedented in terms of its scale, scope, and global rate of adoption. For example, it took almost 40 years for radio to reach an audience of 50 million, and 15 years for television to do the same—but the Worldwide Web (WWW) achieved this goal in just a little more than three years from inception (Naughton, 1999). The proliferation of content has been staggering. There are well over 100 million Web sites on the Internet, and growth reached record highs in 2006, when the Web gained 30.9 million sites during the course of the year (Netcraft, 2007). On average, Internet Protocol (IP) traffic has been growing at approximately 1,000 percent a year, which compares to a rise of just 10 percent a year in traffic on the Public Switched Telephone Network (PSTN). If the demand for bandwidth can be met by new technologies, then IP traffic should easily surpass PSTN traffic, with much of the growth being accounted for through "e-commerce" (i.e., trade that occurs over the Internet) (UNDP, 2001: 36). The global commercial backbone services and network services industry sales were estimated to account for an astonishing $1.3 trillion in 2004 (Chadwick, 2006: 213).

The Internet has excited fevered speculation as to its revolutionary potential, owing to its unique, intrinsic features. It is distinct in that it is a matrix of networks based on a "many-to-many" model of information distribution, as opposed to the "one-to-many" structure of mass media. The continuing increase in computing capacity permits information exchange at dizzying volume and velocity. It is an eminently flexible medium, able to support any application and transmit any kind of data, whether text, images, or sound. As a result, media production and distribution has undergone a process of rapid and radical decentralization. Sites containing user-generated content, such as YouTube and MySpace, are now amongst the most visited domains on the Web. Peer-to-peer file-sharing sites, where media files can be uploaded and accessed amongst a community of users, have attracted devoted audiences. This democratization of the media has already resulted in severe losses by media producers and outlets that were previously entrenched in an oligopolistic market position. For example, the profitability of the record industry has been gravely damaged by the rise of illegal file-sharing music sites. Similar challenges face the mainstream news industry. Large corporations no longer have the exclusive privilege of transnational publication and product promotion. For a small outlay and with a modicum of technical knowledge, people can set up their own Web site or blog and potentially access a global audience of millions. An online presence enhances the accessibility of independent media outlets and grants them a greater reach than ever before. Amongst certain sections of society, such as young Westerners, the implications in terms of news consumption habits have been profound. For example, a recent study of the 18–34 age group in the United States found that 44 percent used the Internet at least once a day for news, compared to just 19 percent who bought a daily newspaper on a regular basis. Further, 39 percent expected to use the Internet more in the next three years, versus 8 percent who expected to make greater use of newspapers. The report speculated that these findings are likely to partly reflect the declining levels of trust that young people have in traditional mass media, noting that only 9 percent of respondents would describe print news as "trustworthy" (Brown, 2005). The survey precipitated Rupert Murdoch to deliver the following warning to a gathering of the American Society of Newspapers Editors

> They [young people] don't want to rely on the morning paper for their up-to-date information. They don't want to rely on a god-like figure from above to tell them what's important. And to carry the religion

analogy a bit further, they certainly don't want news presented as gospel…They want control over their media, instead of being controlled by it. (Murdoch, 2005)

As a self-confessed "digital immigrant," slow to appreciate the ramifications of the Internet revolution, Murdoch was belatedly recognizing the attitudinal shifts that may be attributed to distributional changes in media. Hitherto, public debate has been largely channeled through a limited range of mass media outlets, whereas cyberspace is an infinitely more heterogeneous discursive environment. The coherence of mass-mediated, national public spheres contrasts sharply with the Internet's hyperlinked, interactive structure, which creates a highly complex web of overlapping discussion forums on every conceivable topic. Neither do the same logistical constraints of mass communication apply. The Internet can transcend physical obstacles, empowering those disenfranchised by geography and facilitating deliberative exchanges outside of the nation-state context. In short, it affords unprecedented potential for interactivity and global interconnectivity. Hyperbolic speculation about the deliberative opportunities of the medium continues to thrive, and can seem seductive. But optimism about the Internet's public sphere potential should be tempered by recognizing the extent to which corporate dominance has been replicated online.

The media conglomerates have altered their modus operandi in response to the rise of the Internet, and in some respects they have been successful in maintaining their hegemonic market position. For instance, after years of declining CD sales, the Recording Industry of America (RIAA) have recently announced that illegal file-sharing has been "contained." An industry-led campaign to clamp down on music piracy, which included sponsoring the surveillance and prosecution of 18,000 individual consumers, has achieved notable success. And the increase in sales of legal downloads in 2006—some 77 percent— more than compensated for the 3 percent decline in album sales in the same period (Graham, 2006). In essence, the corporations have turned the tables on the cyber-pirates, by using the Internet to police and prosecute digital copyright infringement. The industry has also benefited from aping the behavior of online music communities. Free file-sharing and "viral marketing" PR campaigns have been used to promote industry acts. This new type of "stealth" PR promotion manufactures the appearance of a groundswell of support. Marketers infiltrate chatrooms, send mass e-mails, and post messages on blogs and bulletin boards. The artificial "buzz" can then be publicized through mainstream outlets as "evidence" of public interest. Corporations have quickly become adept at finding ways to harness the Internet revolution to suit their commercial interests.

In the aforementioned speech, Murdoch forecast ways in which News Corporation could adopt a similar process of assimilation. He argued that the news industry should not perceive the Internet solely as a threat—providing they adapt to the changing behavior of the consumer, it also offers an unrivalled opportunity to increase advertising revenue. Murdoch advocated a more Web-centered, consumer-focused approach to news production, and predicted the economic benefits that could result:

> ...the [I]nternet allows us to be more granular in our advertising, targeting potential consumers based on where they've surfed and what products they've bought. The ability to more precisely target customers using technology-powered forms of advertising represents a great opportunity for us to maintain and even grow market share and is clearly the future...(Murdoch, 2005)

Subsequently, News Corporation bought Intermix Media, the owner of MySpace, for $580 million. The site contains detailed personal information about the users, a large proportion of who are young and affluent. The commercial desirability of this data is obvious.

These examples are representative of a broader trend that is apparent in numerous sectors and across the globe. New media evidently has the capacity to extend the disciplinary influence of the market; to enhance corporate opportunities for manipulation and domination. ICTs are important arsenals in corporate surveillance strategies. Companies routinely collate commercially useful data by a variety of sophisticated and surreptitious devices, such as tracking online behavior, which enables them to build customer profiles. As Murdoch suggests, marketing campaigns can then to be targeted to the consumer demographic with greater effect. As a result, the Internet experience can be considerably impaired for many who are subjected to intrusive "pop-up" adverts, or bombarded with "spam" e-mails. Without sufficient safeguards in place, this type of advertising can make efficient communication and Web-browsing impossible. Furthermore, the Cyber Security Industry Alliance has reported a high level of anxiety amongst the American public over privacy and security issues, and suggested that their findings indicate that "many will not participate" as a result (CSIA, 2006). Thus, the commercial colonization of the Web may limit the opportunities for critical discursive spaces to flourish.

Unfortunately the commercial exploitation of Internet users is growing apace in a weak regulatory environment (Lessig, 1999). Corporate dominance is also evident in Internet software itself. The future development of Internet applications has largely become the domain of big business. Microsoft has held a long-held position of unparalleled dominance as supplier of Internet software. For example, Microsoft's Internet Explorer commands 85.81 percent of the global usage share (although it is facing a growing challenge from the open-source browser, Mozilla Firefox, which has a 11.69 percent share) (OneStat, 2007). Internet Explorer is often already installed on new computers, and hence gives the company a considerable advantage in the marketplace. Microsoft has also produced programs that automatically save a number of Web sites in the "favorites" section of the browser. Prompted by such developments, the U.S. government took legal action against Microsoft as an illegal monopoly. Following four years of legal battle, Judge Thomas Jackson ruled in 1999 that Microsoft had used its monopoly power with Windows to harm consumers, computer makers and other companies. It was proposed that the company should be split in half, but the appeals court eventually overturned the ruling, and the government agreed to a settlement (Antitrust Division). However, controversy about Microsoft's monopoly position continues (Chin, 2005). The EU Commission found the company guilty of abusing its market dominance, and issued a fine of €495m in 2004. The commission was forced to criticize Microsoft again in 2007 for not altering its behavior, the exceptional nature of which was underscored by a spokesperson's comment that "[i]n the fifty years of European antitrust policy, it's the first time we've been confronted with a company that has failed to comply with an anti-trust ruling" (BBC, 2007). Microsoft's defiant response suggests that the commission will continue to struggle to curtail such abusive practices.

Research regarding the patterns of ownership in the rapidly evolving communication industry is patchy. The complexities of the market are such that, as Slevin observes, "it is unlikely that our understanding of these developments will be ever more than partial" (Slevin, 2000: 39). But an increasingly evident trend is that smaller companies are becoming rarer, as large corporations continue to expand by swallowing up their minor competitors. An example is WorldCom, whose acquisitions in recent years has meant that it carries more than half of all Internet backbone traffic, controls more than half of all direct connections to the Internet, and leases line-capacity to two-thirds of all Internet service providers (ISPs) (34). The conglomeration of new media has conferred significant strategic power to a select few organizations.

From the perspective of public sphere theory, the oligopolistic media market is alarming in several respects. First, conditions of monopoly could allow companies to demand license fees for the use of their products. This could further restrict opportunities for Internet access for those on low-incomes (Gimmler, 2001: 34). Second, companies may be able to pursue discriminatory practices against certain users, for example by charging ISPs prohibitive amounts for line-rental. Third, if powerful corporations monopolize the design of applications and control of the interfaces, they will have considerable influence over information content. Not only does this maximize the potential for the commercial exploitation of users, but also the opportunities for corporate actors to manipulate and distort publicity. These fears are manifested by so-called filtering technologies, which can significantly restrict information access and freedom of expression.

4.2.1 Filtering Technologies

Censorship online is mainly carried out through Internet content filtering. Many filtering systems are marketed to parents and schools as "safe portals" for juvenile Internet usage (e.g., NetNanny.com, Cyberpatrol.com, Cybersitter.com). Certainly filters are valuable in restricting children's access to unsuitable material, or to chatrooms where they may be exposed to pedophile "grooming" techniques. They can also be useful in limiting a user's vulnerability to computer viruses. However, the reasons for installing filtering programs range from benign concerns to malign motivations. The technologies are widely used to "spy" on Internet behavior, to monitor personal communications, and to prevent open debate. For example, employers may wish to control Web sites accessed by their employees, and check on the content of their e-mails. Suspicious lovers may want to keep tabs on their errant partners. Even more sinisterly, governments frequently use filtering to clamp down on internal dissent.

Deibert and Villeneuve (2004) specify two main types of filtering techniques. First, "blocking" can be used to deny access to certain IP addresses or port numbers, that host material deemed offensive, such as hardcore pornography. Access can be restricted to a limited number of approved sites, or alternatively Web-surfing can be unregulated outside of a "black list" of undesirable sites. Institutional use of "blocking" filters enables the installment of "firewalls" that prevent access to certain sites (e.g., by an employer or state government). "Content analysis" is a more sophisticated form of filtering, where access to information is controlled based upon the textual and/or graphic content of the site. The filtering system can be programmed with criteria, usually based on keywords related to the offending topic. The user is either denied access to the page or the taboo words are replaced automatically with alternative signs, such as "xxx."

The invention of the Platform for Internet Content Selection (PICS) was highly significant in the development of rating and filtering technologies. It assigned a series of electronic labels to Web sites that conveyed the characteristics of the content. Some labels serve to indicate the presence of adult material; others contain information about the Web site's policies on the use or resale of personal data to third parties. PICS enabled rating programs to access such labels, so that users can set a "content adviser" based on filtering criteria that regulate access to offensive material. One of the most popular rating systems was developed by the Recreational Software Advisory Council (RSACi), eventually integrated into Microsoft's Internet Explorer. It attracted controversy for several reasons (Sobel, 1999). Despite the council claiming to be independent, it had received support from a number of companies including CompuServe PointCast, Dell, and Disney Online. The possible effects of these linkages could not be fully scrutinized because of the limited accountability and transparency of RSCAi programming. The reasons behind the assignment of certain PICS labels to Web sites and servers did not have to be publicly justified. Concerns were raised that in the absence of an effective system of supervisory regulation, the filtering system could be used to assign negative labels to Web sites from its political opponents or commercial competitors.

Possible malicious intent was not the only cause for concern. Filtering is a crude means of censorship, and has inaccurate and unintended effects (EPIC, 1997). For example, RSACi's system had a facility that allows all non-RSAC rated Web sites to be blocked. The council recommended that users choose this option, since new sites are created every hour, and many will not be rated. However, this meant that Web sites would be blocked for no better reason than they had not yet been assigned a PICS label (ibid.). In addition, "taboo" words programmed in the filtering criteria can have alternative meanings or inferences. Hence, gratuitous and offensive material could be debarred together with valid and responsible contributions to public debate. For example, a system that filters out sexual content with the intent to restrict access to salacious pornographic material may also have the effect of blocking serious sites about sexual health. A system that filters out content on sexual violence may block sites that include reports of war crimes. Obviously the meaning and value of a text cannot be adequately derived on the basis of words divorced from their context. Weinberg expands further:

> [RSACi] classifies sexually explicit speech without regard to its educational value or crass commercialism...A typical home user, running Microsoft Internet Explorer set to filter using RSACi tags, will have a browser configured to accept duly rated mass-market speech from large entertainment corporations, but to block out a substantial amount of quirky, vibrant individual speech from unrated (but child-suitable) sites. This prospect is disturbing. (Weinburg, 1997: 455)

Microsoft has now developed its own content adviser and RSACi has evolved into the Family Online Safety Institute. However, amongst free speech advocates, the same misgivings about filtering technologies remain, even in instances where filters have been installed with the user's consent. Users may subscribe to ISPs that use filters without being fully aware of the facility or of its ramifications.

Organizations as Cyber-Rights and Cyber-Liberties and the Electronic Frontier Foundation have led strident calls for careful monitoring, greater accountability, and transparency of the practices of companies involved in filtering and rating systems.[2] Human Rights Watch is a leading participant in this campaign and produces regular reports monitoring freedom of expression on the Internet (Human Rights Watch, 2007). These reports focus on the activities of filtering companies as well as those of censorious governments. Their recent research also has explored the implications of state-controlled or state-influenced ISPs in Tunisia, Iran, and Bahrain that filter Web sites containing political or human rights criticism of their governments (Reporters Without Borders, 2007).[3] As will be seen, Internet censorship and surveillance does not just empower corporate actors, but also increases the disciplinary and repressive capacities of states over citizens.

4.2.2   State Censorship and Surveillance

States from all regions of the world rely on ICT for the most basic functions of governance. From data-processing to multilateral communications, ICT are an indispensable part of modern administration. However, they also present intractable policy problems, facilitating crimes such as identity theft and "cyberterrorism." They also permit citizens to disseminate and access material that governments regard as morally corrupt and/or politically subversive. Hence, as Internet usage proliferated during the 1990s, so did regulatory constraints around the globe.

States have always pursued forms of censorship, and authoritarian regimes are usually the most robust perpetrators. More than a 100 years ago, for example, the import of typewriters was banned in Turkey, reflecting official disquietude with the democratization of print. The authorities feared incendiary pamphlets that promoted rebellion and dissension would be produced. Likewise, Western radio broadcasts were routinely jammed by the Soviet Union during the cold war. Similarly, today the North Korean government exerts tight controls over the access of their citizens to typewriters, photocopiers, and radios. The emergence of Internet censorship across the globe was inevitable.

The Internet's unique structure means that it has a latent potential to circumvent centralized management (and thus complicate efforts at censorship). Such decentralized global information flows are exceptionally difficult for governments to control. Thus to some extent, the Internet has actually hindered the prosecution of censorship, by facilitating communications in dictatorial regimes. For example, Reporters Without Borders have made regular use of the Web-phone service Skype to communicate with sources in authoritarian states, as conversations are automatically encrypted, ensuring a high degree of privacy. There are other ways in which citizens in censorious countries can thwart restrictive laws, such as redirecting the information they send and retrieve to a proxy server, which helps to protect their identity. Indeed, Reporters Without Borders (2005) have a downloadable handbook for cyber-dissidents that gives extensive advice on the safest means of circumventing censorship controls. Also, when sites have been banned, users can simply disperse the information to other sites. For example, when Canadian universities banned a site that published details of a legal trial that the judge had deemed unfit for public consumption, users diverted the information to an alternative address (Shade, 1996). John Gilmore, cofounder of the Electronic Frontier Foundation, describes the problem of regulation thus, "[t]he [I]nternet treats censorship as system damage and routes round it" (cited in Corn-Revere, 2002: 13).

The governmental dilemma has been compounded by the exponential growth in the speed and volume of information traffic. The unremitting deluge means that attempts at control and surveillance are likely to have a limited reach, even those that are well funded and have a high public profile. Thus vulnerable human rights defenders can be shielded by the "safety of numbers." Unfortunately human rights abusers are also sheltered by the crowd. A good example here is the international police investigation, Operation Ore, launched after raids on the offices of a Texan Web site that was distributing obscene images of children. It was the largest ever child protection investigation, resulting in almost 1,500 convictions in the United Kingdom alone. Officers in the Unites States identified 35,000 customers of the site (Cobain, 2006). Yet children's experts commonly agree that this represented merely a small proportion of those likely to regularly access child pornography in both countries.

However, there is conflicting evidence to suggest that the Internet can be better adapted for the purposes of control than can older forms of technology such as telephone, fax, or print. If sufficient resources and political will can be mobilized, the software exists for detailed surveillance across large sections of the population. For example, following the 9/11 attacks, there has been a concerted push from Western governments for greater access to personal electronic data and communications. Ostensibly for purposes of criminal investigation, the measures often alarmingly encroach upon civil liberties. The USA Patriot Act was passed with minimal debate only 45 days after the attacks, which consolidated the FBI's authority to install surveillance software to monitor e-mail content and store details of Internet behavior by those suspected to be in contact with a hostile country. These decisions are only subject to review by a secret tribunal. The act also made it easier for ISPs to share details of Internet activity with the authorities. The American Civil Liberties Union has been at the forefront of a campaign for legislative reform, arguing that the act has had a "profound chilling effect on public discourse...[people] inevitably feel less comfortable saying what they think, especially if what they think is not what the government wants them to think" (ACLU). Similar steps have been taken in the EU, where Article 15.1 of the 1997 directive on confidentiality and privacy was amended in 2002 to oblige ISPs and phone companies to retain all records of Internet activity and e-mails for police and judicial access. Indeed, despite the medium's inherent flexibility, and the sheer volume of information traffic, Internet censorship is well-established and becoming more prevalent. The following brief survey of some of the most notorious examples of global Internet censorship illustrates how governments have become increasingly adept at control and surveillance.

China is the most egregious example of Internet repression, hosting more "cyber-dissidents" in jail than any other state in the world. Amnesty International has exposed an ongoing campaign by the Chinese government to suppress online dissent, claiming that

54 people have been arrested from December 2003 to February 2004 for disseminating their beliefs through the Internet—an increase of 60 percent on the previous year. The crimes that they were alleged to have perpetrated include organizing online political petitions, expressing support for the outlawed Falun Gong movement, and for spreading "rumors" about AIDS and SARS. It was reported that the prisoners face maximum sentences of 12 years, that they are subject to torture, and that four died in detention (Amnesty International, 2004; 2006). These arrests reflect the Chinese government's concern with the rapid growth in Internet users, currently estimated at 111 million (ibid. 2006: 16). A 30,000-strong "Internet police force," has been established to monitor chatrooms and Web sites. Filtering technologies are habitually used at public access facilities to block access to Web sites such as the BBC. In addition, search engines have been filtered so that no results are returned for searches such as "human rights," "democracy," or "Falun Gong" (Kalathil and Boas, 2001, 2003: 27; OpenNet Initiative, 2005a; Zittrain and Edelman, 2003b). Internet cafes and service providers are under increasing pressure to assist in the government's actions. Tens of thousands have been shut down by the authorities for not fully installing the required software filters (Reporters Without Borders, 2004).

Iran is estimated to have the most extensive use of filtering technology after China. Typical targets include the BBC, opposition groups, and sites deemed "immoral" by the religious authorities. Recently efforts have been made to clamp down on dissent and cultural influences from the West by banning some of the most popular sites such as Wikipedia, Amazon, and YouTube (Tait, 2006). The head of the information committee has issued warnings to site owners of the unacceptability of content deemed to pose a threat to national unity, or insulting to religious sensibilities (ibid.). Reporters Without Borders have documented several instances of the intimidation and imprisonment of antiestablishment bloggers (Reporters Without Borders, 2007). Iranian citizens tend to be enthusiastic Internet users and banned material is commonly accessed through proxy servers. The software company, Anonymizer, runs such a proxy system that has been cofinanced by the U.S. government to help Iranian citizens evade censorship. Sadly, this initiative is not fully effective because Anonymizer has its own crude filters installed that block access to sources of legitimate information. In a misguided attempt by the puritanical U.S. government to regulate access to pornographic material, key words such as "boys" and "breasts" are filtered out. These controls block sites with responsible and important content—for example, health information about breast cancer. Ironically, the Iranian government operates its filtering system through commercial software developed by Secure Computing—an American-based company (OpenNet Initiative, 2005b).

Censorship policies are pursued with particular vigor in Arabic countries. For instance, Saudi Arabia's censorship laws are extensive, prohibiting the publication of access of material that includes "anything contrary to the state or the system, news damaging to the Saudi Arabian armed forces, anything damaging to the dignity of the heads of states, any false information ascribed to state officials, subversive ideas and slanderous or libelous material" (Zittrain and Edelman, 2003a). All 30 of the country's ISPs are linked to a ground-floor room at the Riyadh Internet entry portal, where all of the country's Web activity is stored in massive cache files and screened for material deemed to be offensive before it is released to individual users. The central servers are configured to block access to certain sites that might violate "the social, cultural, political, media, economic and religious values of the Kingdom" (Kalathil and Boas, 2003: 113; also see OpenNet Initiative, 2004). Banned sites range from the UK-based Movement for Islamic Reform in Arabia, the International Gay and Lesbian Human Rights Commission, and the companion site of the music magazine Rolling Stone. Likewise, the Syrian government censors the publication or access of online material that it deems to endanger "national unity." This includes statements that are interpreted as being pro-Israeli. Syrian citizens also face jail for sending e-mail to people abroad without correct government authorization. There is only one Internet server in the country, which is run by the government under heavy surveillance (Arabic Network for Human Rights Information, 2006).

In Africa, Zimbabwe has deserved notoriety as a censorious regime. It has a well-developed Internet infrastructure compared to much of the rest of the continent, with 12 large-scale ISPs and an estimated 100,000 users. The Internet was one of the few communication channels accessible to anti-Mugabe activists after independent publishers had been shut down and government control established over radio and television broadcasts. Several subversive Web-based newsletters have circulated in recent years. However, in 2004 the government declared that maintenance of phone-line access for ISPs was conditional on accepting the terms of a contract stating that "the use of the network for anti-national activities will be regarded as an offence punishable under Zimbabwe law" (Meldrum, 2004). The contract also required ISPs to turn over the details of subscribers sending such messages to the police. The Zimbabwe Internet Service Providers Association has protested at the unfeasibility of the policy: "The volume of our traffic makes that impossible. And how would we be able to judge what the government finds objectionable? It would make us the [I]nternet police instead of the [I]nternet providers" (ibid.). The Supreme Court has already declared some aspects of this legislation as unconstitutional. Mugabe was undeterred by this setback, and drafted the Interception of Communications Bill 2006, which will compel operators to intercept and store information at the government's discretion.

Power struggles between ISPs and governments over Internet censorship are not just restricted to dictatorial regimes. Spats are frequent, and often aggravated by the legal ambivalence of the international regulatory framework regarding freedom of online expression. Groundbreaking precedents were set by the long-running dispute between Yahoo! and the French courts over online auctions for Nazi memorabilia. The case was brought by interest groups La Ligue contre le Racisme et l'Antisemitisme (LICRA) and L'Union des Etudiants Juifs de France (UEJF). They argued that the sale of Nazi artifacts violated Article R645–1 of the French Criminal Code, which prohibits the display of any symbol associated with an organization deemed to be criminal. Yahoo! argued in defense that the Internet is globally accessible and companies cannot be subject to the laws of different jurisdictions where their sites may be viewed. Further, as Yahoo! is mainly based in the United States, they claimed a constitutional guarantee of free speech under the First Amendment. In 2000, the court found in favor of the plaintiffs, declaring that French law applied if the material was available within French borders (Corn-Revere, 2002: 4). Yahoo! was directed to do all it reasonably could to identify French IP addresses and to block access to the offending information accordingly. It was also ordered that Yahoo! users who were difficult to locate should be required to declare their nationality when attempting to access such material. Yahoo! was given three months to carry out these changes, and was threatened with a penalty of 100,000 francs for each day of noncompliance. The case was regarded as a highly significant attempt at the extraterritorial application of national law (Greenburg, 2003). It implied the emergence of a transnational legal framework whereby Internet servers could be held accountable for Web content in national courts; regardless of whether the assets of the ISP were sited outside the jurisdiction in question, or the court's decision contravened the laws of their home territories. In other words, Internet servers could be legally obliged to assist in state policies of censorship.

Yahoo! chose to bring a lawsuit against the ruling in the United States, where the District Court for the Northern District of California held that the Yahoo! order could not be enforced in the States. Judge Fogel also rejected the French Court's finding that filtering software could be used. However, the French parties appealed the decision on the basis that Judge Fogel's decision would "give United States Courts worldwide jurisdiction over any non-forum conduct that has the potential of offending local sensibilities" (cited in Corn-Revere, 2002: 11). The Ninth Circuit Court of Appeals issued an ambiguous judgment that asserted jurisdiction over the dispute but also determined that free speech had not been restricted by the French Courts, as Yahoo! had voluntarily removed most of the material. Therefore, the tensions between the global nature of the Internet and the application of national laws are still to be fully resolved.

This issue has progressed following the Council of Europe's (CoE) Convention on Cybercrime 2001. The treaty, also signed by the United States, aims to pursue a common criminal policy against cybercrime, such as defamation and child pornography. The principle established in the Yahoo! case may be put to the test again if the signatory states attempt to apply the protocol extraterritorially to sites based outside of Europe. Indeed, an additional protocol on hate-speech was signed by 12 CoE member states in 2003, which expressly makes cross-border communications of racist or xenophobic material by foreign Web sites illegal (CoE, 2003). Legal uncertainty about Net censorship may have the effect of discouraging users to freely express their views. For example, Tom Krwawecz of Blue Gravity Communications, who was ordered by Italian regulators to withdraw from hosting "blasphemous" sites, protested "[h]ow are we to know what the laws of another country might be" (Thierer, 2002: 2). Likewise, David Farber, former chief technologist at the Federal Communications Commission and the moderator of a popular listserv on technology policy warns: "if this happens too much, and I start getting letters from overseas, it's going to water down my willingness to do things and say things" (ibid.). Perhaps the most insidious threats to freedom of expression are regulations that create a climate of timidity and self-censorship.

Nonetheless, it is not always the case that private actors are pressurized into censorship and customer surveillance under duress. They are often willing collaborators in government human rights abuses—prepared to overlook the ethical consequences of cooperating with repressive regimes in order to promote their economic self-interest. In these cases, a powerful and dangerous nexus forms between tyrannical governments and irresponsible industries with grave implications for freedom of expression. China is the most flagrant example of this unholy alliance. As one of the world's most important emerging info-markets, China promises immense financial rewards for the ICT sector, which is a strong incentive for companies to cooperate with government surveillance and censorship. Yahoo! has provided access to e-mails that resulted in the conviction and subsequent imprisonment of Shi Tao, a Chinese journalist currently serving 10 years for posting information relating to Tiananmen Square on a U.S. pro-democracy Web site (Amnesty International, 2006: 15). Cisco and Sun Microsystems have cooperated closely with the Chinese government to design monitoring technologies for public chatrooms (ibid.: 11). In apparent contradiction to its position on corporate ethics, Google also recently agreed to launch a modified version of its search engine for Chinese use, which filters out sensitive information. Hitherto, Google were famously known for their commitment to full information access (symbolized by their motto "do no evil"). The company itself has publicly recognized that: "...removing search results is inconsistent with Google's mission..." (Watts, 2006). The Chinese search market represented $151 million in 2004 and the online population is predicted to outnumber that of the United States by 2010—it seems that the commercial considerations outweighed the ethical implications (ibid.).

Internet censorship issues will continue to be globally contested between governments, courts, and civil society for many years to come. It is difficult to generalize about the contribution of the Internet to critical publicity in the context of a constantly evolving international situation. The picture is mixed. The Internet evidently provides new channels of cross-border communication, and in certain instances, grants a voice to those otherwise silenced by authoritarian regimes. However, there is evidence that as the Internet has diffused, censorship has escalated and surveillance has intensified, which has had deleterious effects for the norms of publicity.

4.3 DISPARITIES AND INEQUALITIES IN THE INFORMATION AGE

A public sphere must be open and accessible to the widest possible audience. Yet perhaps one of the most distinctive features of the "information age" is the grave extent of global disparities in ownership and access to ICT. These exclusions are structured along the fault-lines of gender, race, social class and education, among others.

They reveal that the so-called information revolution is tightly circumscribed, and also raise concerns about whether the proliferation
of ICTs could exacerbate extant socioeconomic inequalities. The
UN's response to such misgivings has been to announce a Millennium
Declaration about widening worldwide ICT access. The declaration
acknowledges that ICTs are an important tool to achieve the broader
goal of poverty reduction and global development (UN, 2000).

One may be forgiven for thinking that ICT access is an absurdly
superfluous development priority in the context of endemic global
poverty, where huge swathes of humanity lack access to the most basic
facilities. This is an erroneous assumption, as will be discussed in
more detail further on (section 4.3.2). Nonetheless, the sentiment is
understandable, and it is indeed worth reflecting on some appalling
statistics that illustrate the huge gulf between the world's rich and
poor. For instance, of the 4.9 billion people that live in developing
countries, 1.1 billion live on less than $1 a day, 950 million are illiter-
ate, and 2.7 million do not have access to basic sanitation (UNDP,
2004). There are 104 million boys and girls of primary school age
who are not in school (ibid.). Electricity has not reached some 2 billion
people—or a third of the world's population. In 1998, average electricity consumption in South Asia and sub-Saharan Africa was less
than one-tenth of that in OECD countries (Jensen, 2003: 87). The
low ownership of ICT in the developing world is partly because the
technologies are usually dependent on the availability of electricity or
recharging facilities. For example, radio access in rural Africa tends to
be relatively good compared to other media because they are largely
battery-operated (ITU, 2003: 8–9). Increasing ICT diffusion will
demand massive investment in the essential utilities and infrastructure
of the global South.

However, in recent years, there has been gradual narrowing of the
global telephony gap. In 1994, 4 percent of inhabitants in developing
countries owned a fixed telephone line compared to 49 percent in the
developed world; in 2004, those figures stood at 13 and 54 percent
respectively (ITU, 2006c). The rapidity of growth has been startling.
The proportion of subscribers in developing countries did not even
double during 1980 to 1990, but during the next 10 years, the rate of
growth quintupled. Between 2000 and 2005, the number of subscribers tripled once again. Developing countries now account for
60 percent of the world's telephone lines (fixed and mobile), up from
less than 20 percent in 1980 (World Bank, 2006).

Much of the growth can be accounted for by the spread of mobile
telephony, an ideal communication solution for those living in regions

with poor mainline infrastructure (UNCTAD, 2006b). Mobile penetration was just 0.2 percent for developing countries in 1994, but in just 10 years this has increased to 19 percent (ITU, 2006c). This means that whereas fixed line networks have taken over 130 years to reach a billion consumers, mobile telephony will take only 20 years at the current rate of growth to do the same. However, the spread is uneven, as most of the growth in non-OECD countries has occurred in China. Moreover, third generation (3G) mobile phones have hardly penetrated the developing world at all. 3G phones have Internet connectivity, and are proclaimed as the "next telecommunications revolution" (Covell, 1999). They could potentially be extraordinarily valuable to the developing countries, as they involve the greater use of satellite technology to provide multimedia access as a cheaper alternative to telephone and cable based services. An advantage that satellite technology also possesses over "wired" services is the ability to cover the type of lightly populated, rural areas that usually host the poorest peoples on Earth. This is a crucial consideration in addressing inequality in telecommunications access, even in the global North, where an estimated 30 percent of customers are far from population centers (Thussu, 2006: 209). However, nearly all of the 150 million 3G subscribers are located in the developed countries. Just three states account for 100 million of these subscribers: the United States (49.5 million), the Republic of Korea (27.5 million), and Japan (25.7 million) (ITU, 2006a). Thus, statistics must be treated with caution. Apparent success stories can conceal the reality of the deepening "digital divide."

4.3.1 Understanding the Digital Divide

Following the rise of the Internet, the "digital divide" has been endlessly discussed by scholars and politicians alike. It has been defined and applied in varying ways (Gunkel, 2003: 502–504). A simplistic understanding of the term has been popularized by the "Falling through the Net" series of reports provided by the U.S. Department of Commerce's National Telecommunications and Information Administration (NTIA), which defines it as "the divide between those with access to new technologies and those without" (NTIA, 1999: xiii). For the NTIA, the divide has profound implications for U.S. citizens: "To be connected today increasingly means to have access to telephones, computers, *and* the [I]nternet. While these items may not be necessary for survival, arguably in today's emerging digital economy they are necessary for success" (77). The OECD's interpretation of the concept is broader in scope and application. It describes the "digital divide" as an international, multidimensional issue, which involves not just differentials in material access but also in the quality of use. In a comparative analysis of telecommunications structure in OECD and non-OECD countries, the term "digital divide" is defined as "the gap between individuals, households, businesses and geographic areas at different socioeconomic levels with regard to both their opportunities to access information and communications technologies and to their use of the Internet for a wide variety of activities" (OECD, 2000: 5). This wider approach is more satisfactory, as mere physical access to ICT is a necessary, but not sufficient condition for full inclusion. What use, for example, is access to the Internet if one does not have the education or the technical expertise to productively navigate the Web? Unequal *competence* can be as problematic as unequal access (Gandy, 1988).

Competence in this regard is best understood as existing on a continuum, rather than a dichotomy between those who are capable with ICT and those who are not (Cope and Kalantzis, 2000). Warshauer explains that there is considerable variability in the way in which people access and contribute to the new media environment:

> Compare, for example, a professor at UCLA with a high-speed connection in her office, a student in Seoul who occasionally uses a cyber-cafè, and a rural activist in Indonesia who has no computer or phone line but whose colleagues in her women's group download and print out information for her. This example illustrates just three degrees of possible access a person can have to online material. (Warschauer, 2002)

Similarly, statistics on disparities in access to media in the developing world may not adequately reflect the communal sharing of resources: for example, in Africa, 10 people may read the same newspaper or share an Internet account, and a whole village could share a single telephone line or television set (Jensen, 2003: 86). Therefore, although the "digital divide" is a useful descriptive term for global communicative inequalities, it is important to make the caveat that there are complex variations of access, competence and usage of ICTs. The "digital divide" may best be conceptualized as a "social stratification," rather than a simplistic binary definition (ibid.).

The digital divide has become a subject that has gained a high profile on the global political agenda, as the following quote illustrates. It is taken from the G8 Charter on the Global Information

Society signed in Okinawa in July 2000:

> Our vision of an information society is one that better enables people to fulfil their potential and realise their aspirations. To this end we must ensure that [ICT] serves the mutually supportive goals of creating sustainable economic growth, enhancing the public welfare, and fostering social cohesion, and work to fully realise its potential to strengthen democracy, increase transparency and accountability in governance, promote human rights, enhance cultural diversity, and to foster international peace and stability. (G8: 2000)

Despite the seeming consensus at the international level that Internet access is desirable and should be promoted, there has been little positive action taken to reduce the startling disparities between the privileged and the marginalized. There are some encouraging steps, such as in the work of organizations such as the UN ICT Task Force, the G8 "dot.force," the World Summit on the Information Society, the World Bank's Global Development Gateway, and national governments adopting digital divide policies. Also important is the adoption of "social inclusion" policies by the private sector, as well as the emergence of initiatives like "Connect the World," the Stockholm Challenge, and Global Junior Challenge.[4] There is also evidence of community and grassroots initiatives to improve access (Molina, 2003: 145–147). However, these efforts have made only a modest impact, and the huge task of reducing the digital divide will require a more concentrated and coordinated effort on behalf of all of these actors.

I now turn to examine the characteristics of the digital divide, in order to explore the scale of this challenge. It is possible to separate analysis of the digital divide into two categories: the *global digital divide* (the disparity that exists between states, particularly between North and South) and the *intrastate digital divide* (internal state disparities).

4.3.2   Global Digital Divide

The global digital divide is deplorable, but unfortunately deeply rooted. Although almost every country in the world has a direct connection to the Internet, there are only an estimated 840 million people online globally, which represents around 13 percent of the world's population (ITU, 2006c). Dramatically stark differences in opportunity and life expectation exist between the info-rich and info-poor. The sources of inequality are manifold, but income is the main determinant. Certainly, the "information age" sounds like a wildly inappropriate misnomer when one considers that person in a high-income country is 22 times more likely to be an Internet user than someone in a low-income country (UNCTAD, 2006a: xi). Little wonder when the cost of Internet access in a low-income country is 150 times the cost in a high-income country, relative to income (ibid.). And yet 37 percent of the world's population is located in low-income countries (ibid.). The entire African continent—including more than 50 nation-states—only hosts a measly 2.6 percent of Internet users (ITU, 2006c). There are more users in France alone (ITU Statistics). In fact, there is roughly the same amount of Internet users in the G8 countries as in the rest of the world combined (ibid.). The United States is markedly dominant, accounting for a full third of all Internet users worldwide (UNCTAD, 2006b: 5). Whereas in countries such as Bangladesh, a computer is an unimaginable luxury for most, costing eight years average pay (Lucas and Sylla, 2003: 4).

There is an enormous gap in Internet use across countries, even within developed regions. This is demonstrated by the figures on the proportion of households with Internet access—a key indicator for the developed world. For example, Iceland ranks as the nation with the highest Internet penetration level, with 64.79 users per 100 inhabitants, and yet Spain only hosts 19.31 users. Moreover, the Internet penetration levels in Sweden are almost twice as high as they are in France (Chadwick, 2006: 61). Of course, the majority of users in developing nations do not have household access to the Internet; instead they rely on relatives, friends, work, school, or public places such as Internet cafes. It is therefore more pertinent to study community-access facilities when evaluating Internet access for the underdeveloped world. Making international comparisons with this data can be quite striking. For example, 19 percent of primary and secondary schools have Internet access in Mongolia, compared to 15 percent in Malaysia, and only 1 percent in Malawi. At the other end of the scale, 99 percent of Norway's schools are online (World Bank, 2006). Disparities in the level of e-commerce are also vast, as signified by the uneven global spread of secure servers. While developed countries have 300 servers per million inhabitants, developing countries have less than two. In fact, Canada has more secure servers than the rest of the developing world combined (ibid.).

The global digital divide has an added dimension in terms of quality of provision. In low-income countries, broadband availability is sparse, and the reliability and speed of dial-up are often compromised by poor infrastructure. Sometimes even simple Web-browsing can be impossible in this context (UNCTAD, 2006a: 9). An increasing amount of digital content and services requires broadband, which puts full access out of reach for those without high-speed connectivity. Broadband also has critical applications in terms of e-government, e-learning, and e-commerce. Currently these opportunities are largely the preserve of developed states, which host the majority of the world's broadband users. The Asian region sets the benchmarks for Internet access, usage as well as connection speed. For example, South Korea is the most intensive Internet-using population in the world, recording the highest average rates of usage per month. It also has the highest rate of broadband access, mainly owing to large-scale public investment in telecoms infrastructure (Doward, 2006a). Japan is not far behind, with one household in three having a broadband connection (Foster, 2006). Africa is at the other end of the spectrum in the global bandwidth gulf. The entire continent only hosts 0.1 percent of all broadband subscribers (ITU, 2006c). Perhaps no other statistic more effectively illustrates the enormity of global communicative disparities.

A notable characteristic of the Internet that is sometimes thought to be significant in explaining differential access is the predominance of English. This reflects the Internet's development in the United States, and subsequent fast growth in the English-speaking world. It can be argued that even today, hardware and software companies reinforce the language bias of the Internet by producing computers with operating systems and keyboards that discriminate against non-Roman languages. As a recent UNESCO report noted, reliable statistics on the linguistic diversity of the Internet are scarce. However, an Internet sampling study by O'Neill et al. found that English was overwhelmingly dominant, representing 72 percent of the Web pages surveyed (UNESCO Institute of Statistics, 2005). Chinese accounted for a mere 2 percent, even though Chinese speakers are fast increasing. English predominance persists despite the fact that numbers of non-English speakers online far outweigh English native speakers. Non-English speakers account for 64.2 percent of the world online population, and for 5,822 million of the global population, as compared to native English speakers, who account for just 35.8 percent of the world's online population, and for 508 million of global population (Global Reach). In addition, in a sample of 156 multilingual sites, the aforementioned study found that all provided English translations, but less than a third offered French, German, Italian, or Spanish. This despite the fact that 87 percent of the sites were located outside of the Anglophone world (UNESCO Institute of Statistics, 2005). Thus, multilingualism online appears to acknowledge and even reinforce English dominance. Certainly a large proportion of

English-language sites are based in non-English-speaking countries. In some developing countries, this may be due to the fact that not many other local speakers are online, and there is also an incentive to use English to try and garner an international audience (UNESCO Institute of Statistics, 2005). It can be expected that English will remain disproportionately high for some time to come, even though the linguistic diversity of users will increase.

There are also indications that the Internet can strengthen other national and regional languages. An example is the diffusion of news. In terms of print and broadcast, it has been the English-language media (more specifically the U.S. media) that have managed to gain a significant worldwide news distribution. However, online directories such as Online Newspapers or World Newspapers Online give easy access to a huge spectrum of newspapers from each continent, the majority of which are non-English.[5] A perfunctory online search will uncover countless live streaming sites, downloadable media, chatrooms, and blogs in gloriously rich linguistic diversity. It is becoming easier for the members of all types of language communities to gain remote access to government information, educational materials, scientific journals, and the digitized collections of major national libraries. Online discussion groups can increase connectivity between members of geographically dispersed communities, serving as a vital means to preserve links between linguistic and cultural diasporas. Nonetheless, it must be acknowledged that dominant and marginalized languages online are a reflection of global economic and information disparities.

In a recent article, Robert Lucas suggests that this information gap between rich and poor countries would only be temporarily significant (Lucas, 2000). He argues that latecomers to industrialization grow faster than earlier developers by a factor proportional to the average income gap between the two groups, as latecomers avoid the costs of technological innovation. Moreover, as industrialization spreads, and world economic growth slows down, Lucas predicts that income gaps will gradually reduce. He envisions that within the space of 100 years, all or most countries could be at similar levels of income.

Lucas's model makes a mistaken assumption that the world's technological stock remains static; or undergoes only a very slow process of change. In reality, technological innovation is constant and rapid. Therefore it is extraordinarily difficult for latecomer countries to catch up to similar levels of affluence and technological proficiency enjoyed by others. As Henry Lucas and Richard Sylla argue in response, network innovations in the financial, transportation, communication, and electrical sectors have historically been repeatedly characterized by disproportional access. They suggest this pattern is being repeated by the Internet:

> Suppose...that the Internet and related IT are really epochal innovations such as those of the British industrial revolution two centuries ago...If so, these new technologies...might well increase inequality in the world for decades, with political and social consequences that do not differ from those that came with inequalities brought by industrialization after 1800. (Lucas and Sylla. 2003: 7)

Using regression analysis of Internet host data, they find that while certain developing countries are increasing their hosts at a high rate, countries with a significant Internet presence still predominate in absolute terms. This suggests that the divide between states regarding Internet participation will continue to grow (14).

Nonetheless, it is worth restating that the divide is complex, which presents a problem for accurate analysis. The divide is not consistent—in some respects, such as telephony, it is narrowing. The World Summit on the Information Society has pioneered a composite measure that reveals some of these dimensions. The "digital opportunity index" (DOI) incorporates indicators on factors such as affordability, fixed and mobile telephone access, household ownership of ICT, connection speeds and patterns of Internet usage. Countries can thus be assessed according to their differential strengths and weaknesses (for example, the rise of mobile density in the developing world can be measured as a distinct advantage). The DOI reveals that although the United States and Europe are leaders in realizing digital opportunity, Latin America and Central Asia are closing the gap owing to major infrastructural investments and sharp rises in mobile and Internet users (ITU, 2006b). Since the DOI was measured in 2001, the countries that have gained the most have been developing nations such as China, India, Russia, and Brazil. The rapid growth of ICT access and usage in these states has propelled them to the top of the opportunity league. But the recent gains of these countries largely represent nothing more than "catch-up." Developed states still enjoy the fastest speeds of connection and the lowest costs, largely as a result of the inherited privilege of possessing the most advanced electrical and technological infrastructure (not to mention the benefits of economies of scale).

In sum, although the inequities of the broader global political economy are evident in the world communication order, it is possible to identify a general trajectory toward wider use. However, the international picture conceals the social constitution of domestic divides.

4.3.3   Intrastate Digital Divides

Information inequalities are not only exhibited *between* countries, as they also exist most profoundly *within* them. For instance, in affluent America, only 70 percent of citizens are online, whereas the sizeable minority of the population that does not has remained stable over recent years (Pew Internet). Low household incomes limit wider expansion. Adults living in households with annual incomes of $30,000 or less are half as likely to go online as those with the highest incomes (ibid.). Recent research has uncovered a number of interesting insights into the characteristics of this "information underclass" in the United States. The Pew Internet and American Life Project found that 22 percent of Americans had never used the Internet or e-mail and do not have household access (Fox, 2005). Several reasons were provided, ranging from lack of interest (32 percent), no access (31 percent), difficulty of gaining access (25 percent), lack of free time (7 percent) and expense of access involved (5 percent) (ibid.). In addition to the "low-income," this "offline" population also largely derives from categories such as the over-65s, African Americans, and the poorly educated (ibid.). Further, it was found that connection speed has introduced a new element into the divide. Bandwidth is now a more important factor in Internet use than the extent of the user's Internet experience, which has previously been one of the most significant predictors of online behavior. Broadband users are far more likely to spend more time on the Internet, and to be more extensive users across a variety of activities from banking to blogging. The effects are so profound that the author of the report suggested conceptualizing the U.S. divide into three tiers: the truly disconnected (22 percent), those with more modest connections, such as dial-up users (40 percent), and the broadband elite, who mainly have the highest socioeconomic status (33 percent) (ibid.).

Intrastate digital divides in both developing and developed countries reveal the gulf between the poorest and the richest is replicated in all kinds of societies. For example, India is home to Bangalore, rated by *Wired* magazine as top eleventh in a chart of global hubs of technological innovation and excellence. Yet it ranks sixty-third in the 2001 UNDP technological achievement index. Although the country has the world's seventh largest number of scientists and engineers, in 1999, mean years of schooling were only 5.1 years and adult illiteracy stood at 44 percent (UNDP, 2001: 38). Bangalore is a grimly appropriate symbol of the social injustices of digital divides: it is an island of information prosperity located in the midst of the masses of the poor and unschooled.

Consider the internal divide of another rising global power—China. The number of Chinese Internet users is growing at a rate as astonishing as the scale of national economic growth. It will likely host the most users in the world within 15 years. Currently, Chinese users are heavily concentrated around the affluent areas of Beijing, Shanghai, Shandong, and Guangdong, with a miniscule amount located in poorer areas such as Tibet (Warschauer, 2003: 61). All but 2 percent of Chinese users have at least two years of college education; naturally this privilege is a rarity amongst the population at large. The Internet will remain out of reach for the foreseeable future for most of China's 1.3 billion citizens.

As these examples reveal, the characteristics of the majority of Internet users are similar worldwide. In this sense, intrastate digital divides mirror the global digital divide, representing a transnational class of the info-rich. In most countries, most Internet users tend to be relatively affluent. The ITU estimates that high-income earners make up over 43 percent of the world's online population. In contrast, low-income earners only represent 1.3 percent of the same (ITU, 2003). Internet users are also predominately male, young, well-educated, and urban. With the exception of the United States and Finland, men are more likely to use the Internet than women in OECD countries. In the EU for example, 38 percent of women regularly use the Internet in contrast to 49 percent of men (UNCTAD, 2006b: 6). In most developing countries, the gender gap is far more substantial. All countries follow a pattern of the younger population being more likely to use the Internet than the old: in Australia, 18–24 year olds are five times more likely to be online than those more than 55 years old. Likewise, in Chile, 74 percent of users are under 35, and in China the share is 84 percent (UNDP, 2001: 40). Users are also usually well-educated. For example, in Chile, 89 percent of Internet users have had tertiary education, in Sri Lanka 65 percent, and in China 70 percent. They also tend to be urban, such as in India, where 1.3 million of the country's 1.4 million total Internet users are concentrated in four states: Delhi, Karnataka, Tamil Nadu, and Maharashtra (particularly Mumbai) (ibid.). As to be expected, the broadband divide correlates well with the urban/rural split. For instance, in the UK, the top 10 regions in the country in terms of broadband density are located in London and the affluent Home

Counties. The 10 with the lowest density are in the poorer, remote regions such as west Somerset, Wales, and Scotland (Doward, 2006b).

The challenges in addressing the digital divide in developed states pales in comparison to those faced by much of the rest of the world. Less developed countries (LDCs) are often plagued by poor infrastructure, low income and literacy levels, and restrictions on freedom of expression and political participation. In countries where citizens struggle for food, water, and shelter, ICT access is a much less immediate need, and so maybe seen by some as an impractical aspiration. Nonetheless, it has been widely recognized that ICT can be crucial in enabling NGOs, governments, and citizens to improve the quality of life (World Summit on the Information Society, 2005). And not merely in economic terms—ICT diffusion has been shown to increase civic engagement, promote social cohesion, provide entertainment, and enrich learning opportunities (ibid.; Norris, 2001). Indeed, a recent OECD study, which compared data across 30 countries, suggested that the educational benefits are striking. For instance, whilst those students with home computer access had a mean score in mathematics of 514 points, those without such access scored only 453 points (OECD, 2006). Moreover, there are multiple signs that activities such as social networking, decision making, and political activism are migrating online. Thus, those on the wrong side of the digital divide are not only disadvantaged globally, but also in relation to their privileged fellow nationals. Effectively, the info-poor will be excluded from realizing their full potential for citizenship.

As the technology continues to develop apace, the gap becomes increasingly difficult to bridge and the social exclusion it engenders becomes increasingly complete. Anssi Vanjoki, the executive vice president of Nokia, alluded to this fear in an interview about the introduction of mobile broadband, the next major stage in Internet development:

> In the mid-1990s, I said that if you don't have a mobile phone you will be making a declaration that you want to be outside organized society...People said I was crazy, but now everybody has a mobile phone. Today, I'm saying that in 10 years' time the same will be true if you don't have the full [I]nternet in your pocket. If you don't, you will be socially incompetent. (Smith, 2006)

What Vanjoki fails to recognize is that although mobile phones have reached saturation points in developed societies, Internet diffusion is more complex, as successful navigation requires literacy and technical skill. Without these skills, info-poor citizens will certainly find it impossible to be a fully effective member of society, but this position will usually not be a matter of choice.

There are encouraging signs that some of the digital disparities are being successfully challenged, particularly in terms of gender. For example, in Thailand, the share of female users jumped from 35 percent in 1999 to 49 percent in the space of a year (UNDP, 2001: 38). In the United States, women under 30 and black women actually outnumber their male counterparts (Fallows, 2005). There have been ambitious initiatives, such as the EU's *Broadband for all* program, which aims to deploy broadband infrastructure, broadband public services, and the promotion of ICT skills by rolling out substantial funds to poorer, remote regions (EU Commission, 2006). Yet there is much further progress still to be made. The problems in the poorest regions of the world cannot be tackled in isolation, but only as part of a comprehensive poverty reduction program, which includes plans for infrastructural improvement and educational provision. The challenges are immense.

4.4   Conclusion

Transborder communicative capacity is a structural precondition for the emergence of transnational public spheres. This requires freedom of expression, access to diverse sources of information, and wide diffusion of access and ownership of ICT. It is not easy to evaluate the impact of the global information revolution on these deliberative prerequisites. The optimist may point to the long-term trend of rising access to communication technology worldwide, both internationally and within developed and developing states. ICT are breaking down temporal and spatial barriers that were historically regarded as insurmountable. This could indicate that the communicative infrastructure for transnational publicity is being put in place.

But the pessimist would counterbalance this assessment by critiquing the entrenched global exclusions to ownership and access to ICT. States have consistently attempted to retain control over the development of ICT, which sometimes has negative implications for wider public access. For instance, the neoliberal international drive to privatize communication networks has been to the detriment of subsidized services to the economically and socially marginalized. In addition, some states use ICT for political propaganda and/or to stifle free and open debate. Another disturbing trend is the huge influence of corporate interests in ICT development. A handful of mainly

Western (U.S.) media corporations wield considerable potential power to both limit ICT access and to distort public debate.

The digital divide is widely recognized as problematic in the context of an emerging information economy. Worldwide, the digital elite are relatively homogenous: they are male high-income earners, well-educated, and often live in urban areas. From a public sphere perspective, this divide is a serious hindrance to maximal participation, although there are indications that inequalities are being reduced in some countries. However, progress is slow, which has attracted much comment from academics and policymakers regarding how it can be accelerated. Richard Joseph argues that the digital divide "will not be understood if it is viewed as purely a technological phenomenon" (Joseph, 2001: 335). Increasing access to information involves wider questions of international economic development, and of supranational regulatory regimes. As Joseph observes: "Property rights, market and regulatory institutions and information infrastructures (in their broadest sense, not just technological) will be crucial...So too will be the informational and political aspects of policy modelling and decision-making in developing countries" (ibid.). One could also add to Joseph's analysis that the implications of policies in developed countries are equally significant, as this chapter has illustrated, the digital divide is also an intrastate problem. Truly inclusive public spheres cannot be realized if the significant impediments to access and participation are not overcome.

In sum, the intrinsic features of ICTs inhere with a capacity to support the reconfiguration of public spheres across state borders. However, the present world communication order is structured by a depressing triumvirate of multiple social exclusions, concentrated media ownership, and growing corporate and state surveillance. Thus, the structural precondition of communicative capacity is only present for privileged sections of world society.

CHAPTER 5

The Rise of Global Governance: Transformations in Sites of Political Authority

Many societies in history have succumbed to the vain belief that they are afflicted with unprecedented troubles. It is claimed that governance has never been more troublesome, security threats are exceptional, and that human relations are in a state of unparalleled decline. But the winds of change that attract such opprobrium are usually less of a spontaneous tempest and more of a slow-brewing storm. Significant structural changes in society tend to be deeply rooted; hence contemporary situations usually have antecedents. Thus, the current hyperbole about the "uniqueness" of our times should advisably be treated with some caution. Yet there is a growing sense that the increasing interconnectivity of the world is of a different order of intensity than witnessed before. "Globalization" is the *nom du jour* for a complex of social transformations, said to encompass elements as diverse as capital, labor, war, disease, ideas, information, images, news, entertainment, drugs, and pollution (Tomlinson, 1999: 165; for different interpretations of this term, see Scholte, 2000: 15–17). National borders are the conventional organizing principle of cultures, economies, and polities, but under current conditions these seem to resemble porous membranes rather than impermeable shells. There have been a host of claims that we are witnessing the "retreat" or "demise" of the state, and that the economic drivers behind neoliberal globalization are instigating the emergence of novel forms of political organization (e.g., Albrow, 1996; Ohmae, 1995; Strange, 1996). Saurin proclaims that globalization marks the end of the state-centric discipline of IR, arguing that: "only by rejecting *a priori* analytical primacy accredited to the state can one begin to approximate a credible explanation of global social change" (Saurin, 1995: 258). It is indisputable that globalizing forces has historical precursors, for example, the telegraph can be thought of as the Internet of its day. What *is* exceptional about the present is that globalizing forces are coalescing in a potent matrix to challenge the status of the nation-state as the site of supreme political authority (Cerny, 1996). This challenge is highly variable in strength and impact depending on the state concerned. Some are better equipped to resist or subvert globalizing pressures than others. However, the general trend is that the state has lost some of its predominance as a site of governance (McGrew, 1998).

These developments are particularly significant for theorizing the historical evolution of public spheres. Habermas' portrayal of national government as the addressee of public sphere debate was predicated on the primacy of the state as a site of governance (Habermas, 1999: 82). This notion is problematized by recent mutations in state identity. Globalization entails transformations in sites of political authority that may constitute a structural precondition for the emergence of transnational public spheres. This chapter considers the contemporary flux in patterns of governance, state identity, and world order, and examines how state sovereignty and authority is being compromised by changes in several domains. It argues that it is possible to identify the rise of multilayered global governance in the international system, and a polyarchic structure of international authority. While states retain highly significant concentrations of political power, their functions are being disaggregated among other authoritative actors.

The argument proceeds as follows. The first section presents an introductory discussion to the definition of globalization and associated challenges to traditional conceptions of state sovereignty and autonomy. The second section dissects the case for state transformation, focusing on three issue-areas with implications for political authority: the rise of global governance, the growth of international law, and the complex identities of citizens. The third section considers how political authority can be reconceptualized in this context, and argues that the emergence of a structure of international authority problematizes traditional conceptions of state sovereignty. It also notes that global governance is being accompanied by a rising popular crisis in authority. The chapter is concluded with the argument that transformations in sites of political authority provide a structural precondition for the emergence of transnational public spheres.

5.1 Defining Globalization

"Globalization" is a fundamentally vague concept, elastic in scope and application. It is variously used to explain anything from climate change to the dirt-cheap pair of jeans at the mall. It has become convenient shorthand for the entire ensemble of contemporary life. Herein lays its weakness as a noun, verb, and heuristic device. It evades easy definition, but as a rule of thumb, globalization can be understood as a marked increase in the intensity of global interconnections that transcends the state (McGrew, 1998: 300).

Unsurprisingly, there are numerous analytical approaches to globalization that diverge on almost every conceivable aspect (Held et al., 1999: 2–9). For example, skeptics such as Hirst and Thompson (1996) dismiss the idea that globalization is historically distinctive. They argue that the era of the classical Gold Standard was characterized by greater international integration that is the case with current regionalized patterns of economic and social exchange. Aspects of globalization certainly have historical precedents, as illustrated by the network of socioeconomic relations that constituted the transatlantic slave trade. Further, globalization is deeply structured by unequal power relations of distant historical origin. It is extraordinarily uneven in impact, often with disproportionately adverse effects for the world's poorest (consider the differential consequences of global warming in the near future—ironically some of the most devastating effects will be visited on underdeveloped societies that have contributed to global carbon emissions the least). Other skeptics portray globalization as nothing more than a repackaging of old forms of Western imperialism, an ideological smokescreen that reifies the structure of capitalist power relations. Further discussion of the nuances of the globalization debate can be found elsewhere (see Held et al., 1999). Suffice to state that, despite the controversies, "globalization" has gained currency because it seems to capture the essence of our epoch. In the words of one of the prime proponents of globalization theory: "the basic fact of linkage to global flows is a—perhaps *the*—central, distinguishing fact of our moment in history" (Ohmae, 1995: 15, original emphasis). Globalization provides a common vocabulary for understanding these contemporary transformations. The problem is how to avoid depicting globalization in an unnecessarily glib way; in other words, finding a definition that can accommodate the salient complexities.

Anthony Giddens characterizes globalization in terms of "time-space distanciation" (Giddens, 1990). He argues that in a world of instantaneous communication and transnational economic networks, location is increasingly less significant to the structuring of social interaction. Social relations are being "disembedded" from their localized context and reconstituted across time and space by global flows of capital, production, trade, and ICT. These phenomena have considerably diminished national autonomy. Giddens sees this process as a consequence of modernity, with recent globalization as the most intense stage, whereby "... larger and larger numbers of people live in circumstances in which disembedded institutions, linking local practices with globalized social relations, organize major aspects of day-to-day life" (79). He condenses the key dynamics of globalization into the phrase: "action at a distance" (Giddens, 1991: 21).

Held et al. find the definition unsatisfactory and explicate the concept with greater precision. They agree that globalization "stretches" social, economic, and political relations across state borders, so that people's lives are entangled with the lives of others in distant locales. But they point out that these causes and effects do not occur at random; rather there are regular patterns of interaction that have intensified over time. These interactions have accelerated as a result of global transport and communication, which promote the flow of capital, people, and ideas. Growing enmeshment produces effects of disproportionate impact; for example, local events can be magnified to have global implications. Held and McGrew therefore contend that full understanding of the expansive nature of globalization requires an appreciation of its "spatio-temporal" dimensions (Held et al., 1999: 15). They propose the following definition:

> a process (or set of processes) which embodies a transformation in the spatial organization of social relations and transactions—assessed in terms of their extensity, intensity, velocity and impact—generating transcontinental or interregional flows and networks of activity, interactions, and the exercise of power. (16)

The reference to "flows" describes the movements of "physical artefacts, people, symbols, tokens and information across space and time," whilst the term "networks" refers to "regularized or patterned interactions between independent agents, nodes of activity, or sites of power" (ibid.). Held and McGrew submit that their definition is superior to that of Giddens, because it is designed to differentiate globalization from other processes such as "localization," "nationalization," "regionalization," and "internationalization." They couch this claim with the caveat that globalization should not be understood in isolation from such processes, as it will often be implicated in a complex way. For instance, regionalization projects can simultaneously contribute to globalizing tendencies, as is arguably the case with the economic integration of the European Union. Regionalization can alternatively be interpreted as a restraint on further globalization. The effects will vary in different domains and can only be determined by issue-specific empirical evidence.

Globalization *in some form* potentially encompasses all areas of life. Perhaps the rubric is a misnomer, as it suggests a homogenous, uniform process. As Guidry et al. suggest, it may be more appropriate to say there are a multitude of globalizations. Of course, the danger of such a pliable concept is that it describes everything, yet explains nothing. But a measure of equivocality is an unavoidable derivative of global perspectives that, in broad brushstrokes, attempt to capture the spirit of an era. And globalization seems to have something of the zeitgeist about it. Held and McGrew's definition is the most successful attempt to date to mitigate ambiguities by introducing vectors whereby global transformations can be measured and mapped.

The hallmark of the globalization literature is the decreasing relevance of territoriality. It is a theme that receives prominence even in some of the more skeptical accounts. Global flows and networks are crisscrossing boundaries and raising questions about whether social and political relations are as compartmentalized as the realist "billiard ball" model of state policies would suggest. Indeed, the globalization debate has disrupted conceptions of political space that lie at the very core of IR. Nowhere is this more apparent than in the debates and contestations surrounding sovereignty.

Traditionally, nation-states have been regarded as the exclusive bearers of sovereign power in world politics. State sovereignty is the central organizing principle of the theory and practice of international relations. A political authority can be understood as sovereign if it possesses the acknowledged right to govern and determine the framework of rules and policies within a given territory (Held, 1995: 100). Sovereignty has an internal dimension, in that sovereign authority overrides all other forms of power in the territory. It also has an external dimension, in that there is no higher authority than the state in the international realm. McGrew notes that state sovereignty should be distinguished from state autonomy, which refers to "the capacity of state managers or agencies to articulate and pursue their policy preferences either in accordance with the combined pressures of domestic and international forces or on occasion in opposition to these combined forces" (McGrew, 1998: 315). Thus, sovereignty entails the legal *entitlement* to rule, whereas autonomy describes the actual *ability* of the state to attain political goals. Whilst distinct, these concepts are nonetheless interrelated.

Sovereignty is an abstract ideal-type, however, and few states have commanded complete sovereign control in their home territories (Holton, 1998: 86–87). Some are better equipped to do so than others, if blessed with the benefits of strong administration, effective enforcement powers, and a compliant citizenry, among other factors. The poorer states of the world often experience severe restrictions on sovereignty (Hoogvelt, 1997). They may grapple with extreme poverty, resource scarcity, and internal conflict, and have to contend with external inference in domestic affairs by the former colonial powers. Moreover, indebted undeveloped states have been obliged to follow the policy prescriptions of the World Bank and IMF, even in the face of widespread popular and political disquiet (Thomas, 2000). The result is that the policy options for governments are significantly curtailed and citizens have been left with minimal democratic control over their destinies (Jackson, 1990). But rich states are also subject to a multitude of transborder problems that circumscribe sovereignty: such as pollution, AIDS, terrorism, and transnational crime. So sovereignty has historically been more complex than tends to be recognized in conventional accounts. *Entitlement* to rule is enshrined by international law. In practice, sovereignty has been defined according to the political realities of the day (Murphy, 1996).

Numerous analysts have suggested that state sovereignty is continuing to be refashioned by the dynamics of globalization (e.g., Ashley, 1988; Camilleri and Falk, 1992; Campbell, 1993; Held, 1991; Walker, 1991, 1993; Weber, 1992). The distinction between the internal and external dimensions of sovereignty is becoming increasingly blurred by social, political, and economic transformations. The global spread of neoliberalism has fostered the "hollowing out" of the state and the devolution of competencies to private actors and international institutions. The erosion of state autonomy complicates the hierarchy of political responsibility. It has resulted in alterations in the national policymaking climate, and hence the circumstances for administration and enforcement. The changes have been accentuated by the implications of growing global interconnectedness. Simply, policy formation and effectual implementation may not be determined by the state alone. If these trends are sufficiently robust to precipitate significant shifts in the locus of power and authority, then the doctrine of state sovereignty should be reassessed (Held, 1993: 238). Globalization may denote a qualitative permutation of the political terrain, where the traditional internal/external dichotomy fails to mirror the reality of the enmeshment of domestic and foreign policy.

Moreover, the notion of the "national interest"—conventionally understood to be the raison d'être of the state—is complicated under conditions of globalization. The orthodox conception is that the state is primarily motivated to protect the security of its borders and defend citizens from external threat. However, in the current juncture it becomes harder to maintain that states will consistently pursue domestic interests against "outsiders." Rather states are increasingly negotiating the balance between national and international interests, and may perceive the two as serving the same purpose (Koenig-Archibugi, 2003). In this context, sovereignty can be understood less as a claim to supreme authority than as "a resource with which to bargain in the context of complex multilateral networks of governance" (Keohane, 1995: 177). Authority on the other hand is being disaggregated among a wide variety of actors. Thus, sovereignty is markedly different in both form and substance to the traditional conception that is implied in the Habermasian public sphere.

5.2 Transformations in State Autonomy, Governance, and Identities

Scanning the global political landscape reveals an abstruse and perplexing vista. State autonomy, governance, and identity seem to be in flux. This could indicate that transformations in sites of political authority are underway, which would constitute a structural precondition for the emergence of transnational public spheres. I now turn to explore these issue-areas in more detail, by examining the rise of global governance, the growth of international law, and the complex identities of citizens.

5.2.1 The Rise of Global Governance

Political power and relations have been stretched across space and time, beyond the confines of territorial boundaries. The nation-state remains the main concentration of power and authority, but the global policy environment appears qualitatively different from the conditions that pertained decades ago (Cohen, 2001). States are supplemented in the international system by a variety of other organizations, from supranational bodies to transnational social movements, which participate extensively in world politics (Woods, 2003). Increased cross-border exchange in multiple domains has rendered concepts of specifically national economies, polities, and cultures ever more redundant. These developments challenge the conception of the state as a unitary, coherent actor. Even noted "skeptics" of the globalization thesis such as Hirst and Thompson, proclaim that: "There can be no doubt that the era in which politics can be conceived almost exclusively in terms of processes within nation-states and their external billiard ball interactions is passing. Politics is becoming more polycentric, with states as merely one level in a complex system of overlapping and often competing agencies of governance" (Hirst and Thompson, 1996: 184).

Consider the extent to which the international system is characterized by routine multilateral cooperation. Recent years have witnessed much closer collaboration between states in the prosecution of international terrorism, including diplomacy, intelligence-sharing, law enforcement, and military operations. Sometimes states unify to promote a social cause: for example, Germany, Denmark, and Switzerland joined Greenpeace protesters in criticizing Shell over the Brent Spar case (Tsoukas, 1999). Also many states invite consultations with NGOs, think tanks, and epistemic communities when designing and implementing environmental policies. Moreover, state policy is often geared toward the interests of global capital. The Washington Consensus impels many countries to liberalize their economies and reduce corporate tax rates to attract foreign direct investment (Tanzi, 1995). The influence of the global financial market and TNCs is sometimes seen to predominate over more parochial economic concerns (Scholte, 2000: 139).

This picture is a poor fit with notions of international anarchy. "Anarchy" suggests disorder; an undisciplined array of independent actors with little regard for rules and regulations. Yet the international system is evidently *governed*, insofar as there are observed norms and patterns of order, even if there is not a centralized body of world *government*. This growing institutionalization of world politics is commonly termed *global governance* (Rosenau and Czempiel, 1992). Väyrynen defines it as "collective actions to establish international institutions and norms to cope with the causes and consequences of adverse supranational, transnational, or national problems" (Väyrynen, 1999: 25). The activities of state governments are included in the concept of "governance," but so is the variety of other command mechanisms through which global life is organized. The spectrum can be conceived as consisting of three main dimensions.

The first dimension is international institutions, which have been a feature of world politics since the middle of the nineteenth century. They have hugely increased in number over the past few decades, reflecting the postwar growth in communications, trade, foreign direct investment (FDI), and cultural flows (Held and McGrew, 2003: 1). They participate in policy-areas from global health (World Health Organization, WHO), to food security (Food and Agriculture Organization, FAO) to the monitoring of weapons proliferation

(International Atomic Energy Agency, IAEA). Some are heavily oriented around states, such as the World Economic Forum (WEF) and the Group of 8 (G8); others are a combination of state and private actors, such as the International Labor Organization (ILO). There has been a general trend toward an expansion in the remit of key international institutions. For example, the Bretton Woods agencies are instrumental in shaping the economic policies of indebted states, and in mediating strategies to address regional economic crises. Organizations such as the UN or NATO have competencies for resource distribution and conflict resolution, both between and within states (Woods, 2001). Sometimes bodies have been established specifically to adjudicate after civil wars, such as the International Tribunals for Rwanda and the former Yugoslavia. Likewise, the WTO tribunal can authoritatively arbitrate on international trade conflicts. These institutions attract a lot of political heat because they are perceived to make sensitive incursions into state sovereignty; but not all are as controversial. For example, the International Civil Aviation Organization (ICAO) is generally seen as essential for safe global air navigation. Indeed, in an increasingly interconnected and interdependent world, governments may be unable to deliver in core policy domains in the absence of significant international collaboration. Without international organizations to host such collaboration, "it is impossible to imagine contemporary international life" (Schermers and Blokker, 1995: 3).

The second dimension is transnational policy networks between state representatives and official bodies such as regulatory agencies, the police, and judiciary. Their existence indicates the frequent need of states to formulate and implement policy holistically in order to improve the chances of effective problem-solving. For instance, it is difficult to adequately secure borders against illegal immigrants unless synergetic policing and intelligence strategies are pursued between home and host states. A network of consultation and cooperation thus emerges, whereby specific agencies of state interact with their international counterparts to construct a governance framework. This form of "transgovernmentalism" is well-established in areas of routine administration, such as safety regulations, the environment, antitrust legislation, and so on. According to Slaughter, such "transgovernmentalism is rapidly becoming the most widespread and effective mode of international governance" (Slaughter, 1997: 185).

The third dimension is non-state actors, which include private bodies as well as transnational social movements. Private-authority mechanisms have been promoted by acolytes of neoliberalism as efficacious and market-friendly, and have consequently rapidly expanded in the past 30 years. Examples include the International Chamber of Commerce (ICC), mercenary companies such as Executive Outcomes and Sandline, and the not-for-profit Internet Corporation for Assigned Names and Numbers (ICANN), which monopolizes global administration of Internet domain names. NGOs have mushroomed from 31,085 in 1994 to 61,176 in 2007 (UIA, 2007). NGOs based in the global North, such as Oxfam, tend to have the highest profile, but actually the South has experienced the fastest rate of NGO proliferation (Boli and Loya, 1999). NGOs have twin functions in the global governance model: to campaign and exert pressure for political change and to cooperate with governments and international organizations in policy delivery. In the first instance, NGOs have sometimes achieved remarkable success, exemplified by the Campaign for an International Landmine Treaty. In the second instance, NGOs play an integral role in issue-areas from international humanitarian aid to consumer protection, discharging diverse functions that cannot be fulfilled by formal institutions (the composition and impact of social movements are considered in detail in the next chapter). Whether private body or NGO, non-state actors are now an irreplaceable feature of the global political landscape—practically omnipresent in almost every area of policy formulation and implementation. In the words of one observer: "loose alliances of government agencies, international organizations, corporations, and elements of civil society such as nongovernmental organizations, professional associations, or religious groups…join together to achieve what none can accomplish on its own" (Reinicke, 1999: 44).

Global governance resonates with the older concept of "complex interdependence" and regime theory. International regimes are most famously defined in the following terms by Krasner: a set of "implicit and explicit principles, norms, rules and decision-making procedures around which actor's expectations converge in a given area of international relations" (Krasner, 1982: 2). There is a respectable corpus of international regime literature that describes how numerous areas of transnational policy are managed by networks of omnifarious individuals, groups, networks, and organizations (Young, 1989). Regimes institutionalize forms of problem-solving cooperation in a wide spectrum of subject areas, ranging from regulations on access to genetic resources to the legal status of celestial bodies. The extent of institutionalization is context-dependent. Some are centrally administered by an intergovernmental organization and enshrined in international law; others are more flexible, ad hoc arrangements. The membership of regimes also varies, from bilateral arrangements to broad regional partnerships.

For decades, liberal internationalists have countered realist ideas of anarchy by demonstrating how international society is pervaded by all conceivable types of regimes. They may be forgiven for cynically querying to what extent global governance is a departure from this scholarship. The answer is that, in many senses, it is a continuation. Yet global governance is much more than just a synonym for international regimes. Regimes are specific to an issue-area, as described in Krasner's definition. Governance, on the other hand, refers to a broad apparatus of arrangements for a wide range of issue-areas, including (but not restricted to) international regimes. In short, global governance is all-encompassing. It reflects a marked intensification and diversification of multilateral relations from 30 years ago. Above, I delineated different dimensions of governance. The interplay of these dimensions forms a *multilayered* governance mosaic, comprising the supra-state, the regional, the transnational, and the substate (Scholte, 2000: 143–150). So for example, the world economy is jointly managed by the World Bank, the IMF, the WTO, the G8, the Bank of International Settlements, regional bodies like the Association of Southeast Asian Nations (ASEAN), and the European Central Bank. The governance network is interlaced with input from national quangos, devolved and local governments, trade unions, and business councils (Griffin, 2003). The list goes on.

It must be emphasized that global governance has grown organically. There has been no purposeful composition of its present contours by a superior authority. Thus, the players tend not to converge in a harmonious symphony, but rather clash in a chaotic cacophony. There is minimal coordination, as Cable observes: "[i]t is an untidy world with overlapping jurisdictions and competition between different kinds of rules and institutions" (Cable, 1999: 54). With the input of so many varied actors, the structure of international regimes is often highly complex. The "messiness" of the system is often exacerbated by the unsatisfactory outcomes of realpolitik. UN legislation is infused with contradictions and obscure nuances. Take the UN Charter as an example, which is ratified by states as an international treaty. It enshrines the principle of sovereignty, and proclaims the nominal equality of states. The five permanent members of the Security Council (the United Kingdom, the United States, France, Russia, and China) hold exclusive powers and have retained their privileged position since the establishment of the UN. For instance, the General Assembly and the Security Council can only pass resolutions if the five permanent members withhold their right of veto. Likewise, amendments to the charter depend on their consent (Fassbender, 1998: 574). There is another dimension of disparity between state and non-state actors. It is generally accepted that the charter recognizes subjects of international law other than nation-states, such as intergovernmental organizations, cultural groups, minority groups, and individuals (Held, 1995). Despite this, only state governments have participatory rights and hold legislative and executive power. The confusion demonstrates central characteristics of global governance: lack of coherence, competence, and efficacy. Certain governance functions may be outside the ambit of the state, but are not necessarily discharged by other actors particularly well. Moreover, the lines of political accountability are blurred, resulting in a legitimacy deficit. This shall be discussed in more detail below (section 5.3.1).

Naturally the concept of global governance is not uncontroversial, and it has been appropriated by IR scholars from various theoretical traditions. For example, realists maintain that the true operation of power in the international system can only be understood from a state-centric standpoint. In this perspective, compliance is inherently problematic in a world of self-interested states. Clearly it is a truism of world politics that nation-states remain key agents of political authority. Also, the limitations of international law have been thrown into sharp relief by the 2003 Iraq invasion, and the U.S. repudiation of the Kyoto Protocol. Powerful or "rogue" states can choose not to comply if they perceive that their national interests are best served by ignoring their international obligations. In such instances, realism may have explanatory value. However, the intense global controversy that the invasion provoked also indicates that the UN system is widely regarded as legitimate, and that observance of the charter is considered as important in preserving world order. Incompliance does not necessarily equate to lack of authority. This is also a theme to which I will later return (section 5.3).

As for self-interest, the interests of the national political community naturally remain cardinal, and often critical to the shaping of policy. Examples include the popular hysteria surrounding illegal immigration in the global North, which has led to draconian border controls, the discourse of "homeland security" in the United States, and the attempts to protect French language and identity from the onslaught of anglophone culture. Development critics have also charged Northern states with practicing double standards regarding free trade—preaching the virtues of neoliberalism and yet implementing protectionist policies to conserve their home industries. However, the constituency that governments attempt to serve is partly internationalized, which requires that states juggle national, regional, and global interests in their policy strategies (Cerny, 1999). It is an intensely tricky task. It means that the role of the state has not remained static. It has been transformed owing to the ways in which its competencies have been disaggregated across different regimes, bodies, and agencies. Policymaking is more of a fragmented process hosting channels of influence from multiple actors. The state remains strategically important because of its de jure claim to sovereignty, and its de facto powers. Nevertheless, it is not always the principal site of governance (Pierre and Peters, 2000). These changes call for a more nuanced understanding of sovereignty than is permitted by the realist perspective.

Nowhere is this more obvious than the EU, the most ambitious experiment in political and economic integration in the world. Here it seems that theoretical assumptions about the indivisible and territorial aspects of national sovereignty are truly outdated (Keohane and Hoffman, 1990: 10; Wallace, 1994: 20). Supranational governance has been promoted by the European Court of Justice, which has taken an assertive role in establishing the supremacy of European law. It has observed that "by creating a Community of unlimited duration, having its own institutions, its own personality...and, more particularly, real powers stemming from a limitation of sovereignty or transfer of powers from the States to the Community, the member States have limited their sovereign rights" (Mancini, 1990: 180). This has been a freewilled surrender of sovereignty by the European states, in the interests of peace and stability, to protect key industries, and to enhance global economic competitiveness. EU law permeates areas thought to be the most jealously guarded prerogatives of nation-states, such as currency and immigration. The recent expansion of the EU into Central and Eastern Europe, and the ongoing negotiations with Turkey indicates that a variety of states will readily accept compromises to national sovereignty in exchange for the political and economic benefits promised by EU membership.

The EU can be seen to illustrate that sovereignty can actually be enhanced by membership of international institutions, if it better enables a government to meet its policy preferences. All too frequently, states are uniformly cast as passive victims of encroaching globalization. But these forces did not arise of their own accord. States have actively encouraged, indeed are inextricably linked with the nurture of international connections and interdependencies. Thus, sovereign power is not always the victim of externalities, and it is often voluntarily relinquished. Although globalization poses considerable challenges to sovereignty, it is not merely an erosive force. There is a huge diversity of states, and some will be strengthened rather than weakened by globalizing trends

(Giddens, 1990: 731–734). To return to the French language example, feelings of national sentiment are invigorated when it is perceived that the "globalization barbarians are at the gate." There is some comparison here with the rise of political claims across the world for regional and ethnic self-determination. Hence, globalization can engender social integration as well as fragmentation. It is irrefutable that sovereignty is of huge emblematic importance to many decision-makers and citizens. However, I suggest that continuing to place state sovereignty at the centre of our intellectual efforts to comprehend the international system can hinder our perceptions of the patterns of change that are combining to produce world order transformation.

Sovereignty in practice, then, is being challenged by contemporary political and economic trends, which affect the extent of autonomy that national governments can exercise. I want to expand further on how the legal and theoretical aspects of state sovereignty have also been complicated by international law.

5.2.2   The Growth of International Law

International law has established rights, duties, powers, and constraints that transcend the sovereignty of nation-states, constituting an autonomous international legal order. According to conventional wisdom, international law properly exists between nation-states: states are the subjects of law, and citizens are the objects. State sovereignty has traditionally been upheld in the international community by the legal principles of *immunity from jurisdiction* and *immunity of state agencies*. The first prescribes that "no state can be sued in the courts of another state for acts performed in its sovereign capacity" (Cassese, 1988: 150–151). This position had been challenged by developments long before the post–cold war period. A number of landmark agreements have progressively promoted the international legal standing of the individual (Vincent, 1992). Examples include the UN Declaration of Universal Human Rights (UDHR), the European Convention on Human Rights (ECHR) and the UN Convention on the Rights of the Child (one of the most universal human rights agreements, to which nearly every UN member state is party). These statutory instruments establish the rights of individuals that surpass the rights that are guaranteed by the state. Even though states are the only treaty signatories, the key articles permit claims to be made in the name of humanity. Thus, the human rights regime binds states, organizations, peoples, and individual human beings into a common framework (Nickel, 2002).

The 1948 UNDR is widely regarded as the lynchpin of human rights treaties. It stakes its claim to legitimacy in the preamble proclamation: "We, the peoples..." Held argues that this humanitarian stance, and the document's reference to "equal and inalienable rights" establishes something of a blueprint for a cosmopolitan legal order. The international human rights framework has since expanded to incorporate recognition of diverse civil, political, economic, cultural, and social rights. We are daily confronted with evidence in the daily world news bulletins that rights are largely observed in the breach rather than the practice. This does not diminish their symbolic or normative significance. As Held argues: "[h]uman rights entitlements can trump, in principle, the particular claims of national polities; they can set down universal standards against which the strengths and limitations of individual political communities can be judged" (Held, 2003: 315).

Some rights are underpinned by a robust system of enforcement. Indeed, citizens of signatory states to the ECHR can bypass their home governments and directly petition the European Commission on Human Rights, and could then be referred to the European Court of Human Rights. In Goldstein and Ban's analysis, "...few aspects of domestic policymaking pertaining to human rights evade the reach of the Convention and its protocols" (Goldstein and Ban, 2005: 157). The number of applications to the court grew from 1,013 in 1988 to 10,486 in 2001 (156). Europeans within member states of the EU have these rights enhanced by the citizenship status conferred by the Maastricht Treaty. These include the right to free movement, employment, and residence within the community, the right to vote in European Parliament elections and to stand as a prospective candidate. The European Court of Justice can also hear claims brought by citizens who believe that EU law has been contravened or misapplied, once the case has been referred by a lower court.

The second principle on which international law has conventionally rested is the rule of *immunity of state agencies*, which instructs that

> should an individual break the law of another state while acting as an agent of his country of origin and brought before that state's courts, he is not held "guilty" because he did not act as a private individual but as the representative of the state. (Cassese, 1988: 151)

However, human rights law establishes not just rights but the *duties* of individuals, which prescribes that citizens are obliged to disobey state fiat in certain instances. The Nuremberg trials were ground-breaking in this regard, decreeing that individuals are bound to follow international humanitarian rules, even where these may conflict with national law (132). The rulings challenged the superiority of the state over its military—striking directly at the heart of a relationship that is definitive of national sovereignty. Subsequent international law has generally sanctioned the Nuremburg position (Howard et al. 1994). The attempted trials of Pinochet and Milosevic have sought to build upon these foundations, but have been met with mixed success. Nonetheless the possibilities that Blair and Bush might face future legal consequences for the Iraq conflict are being seriously mooted in knowledgeable circles.

Hence, statist immunity principles have been diluted by international courts, often with assistance from national courts and global civil society. Brunkhorst cites the example of the *desaparecidos* in Argentina, which involved local, regional, national, and international legislation. A network of judicial inquiries and cases in different courts were also brought forward: for example, actions were brought in criminal courts in Spain, Switzerland, France, Germany, Italy, Sweden, and a U.S. Civil Court (Brunkhorst, 2002: 683). This was the result of political mobilization by social protest movements like the Argentinean *madres*, who were supported by international associations of legal academics, NGOs such as Amnesty International and Human Rights Watch, and general media publicity (ibid.). Thus, the overlapping networks of legal jurisdiction acted in concert to penalize norm-violations. The example reveals that global governance doesnot have to result in entropy. The multiple players and waves of influence can synchronize to good effect. The situation will vary depending on the context, because like all regimes, the human rights framework is dynamic. It is constantly shifting as a result of new international agreements and the latest adjudications.

Held (2002) sees other distinctly cosmopolitan elements in the emergent apparatus of international law. He suggests that there are international legal bases to scrutinize the nature, form, and operation of state power. In other words, the legitimacy of state governance can be evaluated against international standards. Although the law is far from fully formed on this point, it is notable that, for example, the UDHR asserts that democracy is a "common standard of achievement for all peoples and nations" (UN, 1988: 2, 5). The ECHR is much more specific; indeed, democracy and respect for human rights is a condition of EU membership (Held et al., 1999: 69). Turkey has undertaken an extraordinarily ambitious program of political, economic, and cultural reforms in the hope of gaining admittance to the EU, much to the consternation of traditionalists.

Held is also interested in the cosmopolitan implications of environmental treaties. The principle of the "common heritage of mankind" was established by the Convention on the Moon and Other Celestial Bodies and the Convention on the Law of the Sea. Environmentalists have constantly lobbied to extend this precedent in campaigns for the protection of the "global commons" (Held, 2002). But environmental agreements are always replete with tensions between universal interests and national interests. Take the UN Framework Convention on Climate Change, a crucial step forward in addressing one of the most pressing problems of our age. The preamble acknowledges that "change in the Earth's climate and its adverse effects are the common concern of humankind" and that the "global nature of climate change calls for the widest possible cooperation by all countries" (UN, 1992). However, the emphasis is on "reaffirming the principle of sovereignty of States in international cooperation to address climate change," and in recognizing that states have the "sovereign right to exploit their own resources pursuant to their own environmental and developmental policies" (ibid.). The example reveals that although the international legal order is autonomous, it is not totally self-sufficient: it cannot be separated from the legal order of nation-states (Holton, 1998: 88). Conversely, the national legal order is autonomous but also interwoven with international law. The state-forms but *one* source of legislation and jurisdiction; in addition there are many international counterparts (Rosenau, 2003: 73).

There is one important difference between domestic and international: enforcement. There are few options for international bodies to compel defiant states to participate in the legal process, or to abide by judicial decisions. As Suter observes, UN member states are axiomatically affiliated to the International Court of Justice, but only 65 accept its jurisdiction (Suter, 2005: 131). The fragility of international law has been exhibited by a number of recent events. Iran has engaged in uranium enrichment in violation of UN Security Council demands. To almost blanket international condemnation, the United States has flagrantly flouted the Geneva Conventions by operating the Guantanamo Bay detention camp. Similarly, the increasingly draconian measures taken by European governments in relation to asylum seekers reveal a cynical disregard for the 1951 Refugee Convention. Perhaps the common link between these cases is that each state has perceived a lack of international political will for robust law enforcement. Thus in certain contexts, states can seemingly exploit a certain measure of immunity when they are recalcitrant. There is no over-arching sovereign to consistently prosecute violations. There is no standing body of international law enforcement officers. Prosecution is situational. It depends on the offender concerned, the nature of the offence, and the prevailing international conditions. The state has no competitor as the *main* authority for law enforcement, even when international in origin. For instance, Pinochet was owing to stand trial at a *Chilean* Court (i.e., a national court) before his death cheated his many victims of justice. Even in the EU, the world's most supra-national institution, Community law only comes into effect once ratified by national parliaments and implemented by national courts (Alter, 1996, 1998). It is a pragmatic reconciliation of the tensions associated with the commingled national/international legal orders. Brunkhorst conceives of the result of interpenetration thus: "states have become a community of interpreters of global law, who adjust different legal cultures in association with an international profes-sional class of legal advisers and international lawyers" (Brunkhorst, 2002: 685). The legislative basis of sovereignty is mediated through the case law, the treaties, and the statutory instruments that form the architecture of the global legal order.

Notwithstanding the above caveats regarding the resilience of the state, it can still be maintained that international society has undergone a postwar transformation in terms of its character and its objectives

> ...away from minimalist goals of co-existence toward the creation of rules and institutions that embody notions of shared responsibilities that impinge heavily on the domestic organization of states, that invest indi-viduals and groups within states with rights and duties, and that seek to embody some notion of the planetary good. (Hurrell, 1995: 139)

It could be argued that the cosmopolitanism fostered by international law has contributed to increasing complexities and contradictions in people's identities. This leads to the question of whether the primacy of national identity can be taken for granted when political and legal aspects of sovereignty have been destabilized.

5.2.3   Complex Identities of Citizens

The Westphalian system has historically rested on intersubjective feel-ings of national identity. Since the modern state came into existence, most governments have sought legitimacy by actively cultivating a sense of identity among subject peoples as "citizens." Dominant ideas of citizenship are promoted by the elite as a mechanism for social cohesion. Thus although ostensibly inclusive, the citizenship discourse tends to discriminate in favor of the most powerful groups in society. We have already seen how the public/private dichotomy of citizenship contained a masculinist orientation. Citizenship can be a potent tool of ideological and political repression. Many individuals and social groups have contested the constructs of "nation" and the "state" that constitute their home jurisdiction, and made alternative claims for recognition and self-determination. For example, Basque separatists in Spain demand regional autonomy; and Roman Catholics proclaim a loyalty to the church which supersedes allegiance to their home state. The history of nation-states is littered with examples of similar hostilities that have spilled over into bloody conflict. The relationship of citizens to the state has always been interwoven with power struggles and riddled with ambiguities, insecurities, and controversies.

However, the social bond symbolized by citizenship also partly reflects genuine grassroots sentiment; after all, many states are based on primordial linguistic, cultural, and historical ties. Moreover, most states were born in the struggle for independence or in defensive reaction to warmongering neighbors—and little else bonds together a population more effectively than the presence of external threats. The "imagined community" of the nation-state was partly constructed on the "social glue" of memories and myths, with concomitant understandings of the rights and duties imposed by membership (Anderson, 1991).

It is interesting to consider the process through which the "imagined community" of the nation-state emerged, as media technologies of the time played a critical role. The invention of the printing press precipitated the formation of communication networks throughout the territory. Thus, citizens could establish social and economic relationships with other citizens outside of their immediate vicinity, yet within state borders. As Habermas vividly describes it in *STPS*, print produced an environment in which literary culture and political dialogue could flourish. Further, standardized printed documentation was vital for modern state bureaucracies, because, as Deibert observes, "the preconditions for centralized administrative rule depended not just on ideas, but also, and more crucially, on the technological capacity to carry them out—a distinctly absent feature for most nascent states in medieval Europe" (Deibert, 1997: 87). State action was hereafter based on mass-produced legislative and administrative documents. But print did not just circulate within political communities, but also beyond, which helped to enhance cross-cultural awareness.

For example, the portable medium of the popular novel was widely diffused beyond its country of provenance. These developments progressively undermined the cornerstone of traditional societies: the oral culture. In an influential thesis, Tonnies argues that this transformation represents the erosion of a precious aspect of our common heritage and the human experience (Tonnies, 1957; also see section 6.2).

The introduction of print reminds us that globalizing tendencies originate in developments anterior to digital convergence. But the present situation is qualitatively different from the past. The momentum is far more rapid and extensive. ICT are the engine fuel of the globalization juggernaut; they are deeply implicated in the social, political, and economic transformations that give rise to questions about state sovereignty. One could say that the ancient Silk Road has a virtual equivalent in the Digital Highway—but the latter route permits the *instantaneous* transference of capital, goods, and services. The contemporary world is one where the particularities of locality are continually mediated by transnational communication networks (Street, 2001: 173). Consequently a far greater number are conscious of peoples, societies, and events outside their home territory than ever before. Giddens terms this an "expansion of horizons" (Giddens, 1990: 77). For the citizen, new media opens up a new social and cultural vista that transcends the materiality of everyday experience (e.g., Ott and Rosser, 2000; Bray, 2000; Hill and Sen, 2000; Ferdinand, 2000). It is true that print has a virtual element too, insofar as it enables the circulation of discourse beyond the immediate interpersonal context. But the potential of ICT to liberate discourse from the constraints of time and space is of a different order of magnitude. For many, it is nothing less than a revolution of the "situational geography" of political and social life, since the media increasingly "...make us 'direct' audiences to performances that happen in other places and give us access to audiences that are not physically present" (Meyrowitz, 1985: 7).

The shift was epitomized by the 1989 demonstrations in Tiananmen Square, a watershed in contemporary news coverage that was watched by millions all over the world. Without the presence of the world's television cameras, the massacre would certainly have not have attracted a similar level of international outrage (Calhoun, 1989). More recently, the events of 9/11 had a global audience as collective witness. The attacks and the aftermath were captured at every conceivable angle by ordinary citizens with camcorders and mobile phones, lending a visceral impact to the footage. Consider too the effect of the "shock and awe" campaign in Iraq, which broke new ground as the first war to be broadcast in real time. As the war has progressed, other defining images have become etched on the public consciousness: the prisoner abuse at Abu Ghraib, captured on mobile phones; the sickening footage of hostage beheadings uploaded on terrorist websites. The constant bombardment of live news feed has radically transformed the range of our everyday consciousness. We are immediately cognizant of far distant events in richer and more graphic detail than ever before. This public awareness can be harnessed for direct political effect. For example, the satellite pictures of the hole in the ozone layer over Antarctica were widely recognized to be instrumental in galvanizing public opinion in favor of CFC regulation.

Global media exposure forms a "common ground" of knowledge and experience between peoples whose lives might be divergent in every other aspect. Media publicity can help to generate a sense of commonality among people about world events, based on feelings of empathy and self-identification with social groups. These feelings can destabilize the primacy of national identity and heighten the tensions within the citizenship discourse. Indeed, there has been a noted transformation in the nature of much political struggle since the end of the cold war, known as "recognition politics." A short diversion on this topic will help with conceptualizing the broader challenges to national identity.

Whereas for much of the twentieth century, social conflicts tended to center on socioeconomic reform, they are now increasingly framed by claims for cultural recognition by subordinate groups. The struggle for recognition entails such issues as personal identity, multiculturalism, and ethnic and national self-determination; representing a "paradigm shift" whereby "group identity supplants class interest as the chief medium of political mobilization" (Fraser, 1995: 167). Theoretical inquiries into the significance and legitimacy of recognition-based conflicts have mushroomed in recent years (e.g., Benhabib, 2002; Feldman, 2002; Fraser, 1995; Honneth, 1995; Taylor, 1994; Young, 1990). Habermas has also shown an interest in new social movements as the vanguard of identity politics. He argues that the emergence of activism around issues of equal rights, participation or individual self-realization is a distinct shift from previous forms of collective action, which tended to focus on issues of economic redistribution (Habermas, 1987: 392). In contrast, Fraser and Honneth stress that both forms of conflict are extant, but propose different conceptual approaches. The disputation between the two has shaped the contours of the scholarly debate surrounding contemporary identity-based conflict.

Fraser conceives of a spectrum of ideal-types, ranging from "exploited classes" suffering economic injustice and demanding redistribution to "despised sexualities" experiencing cultural denigration and demanding symbolic recognition (Fraser, 1995: 167). In the middle of the spectrum are groups such as women's movements and racial minorities, which suffer economic injustice *and* cultural denigration. Fraser suggests that the demands of such groups will include claims for remedial measures for both forms of discrimination. Fraser has more recently developed a "bifocal" analysis that is simultaneously attentive to those injustices rooted in class structures and those rooted in institutionally anchored status hierarchies (Fraser and Honneth, 2003: 31). She emphasizes the complexity of oppression, and argues that economic class and social status are interwoven and cannot be treated as analytically distinct forms of justice. The emancipatory antidote requires a combination of recognition *and* redistribution. This leads to the question of how justice is to be applied in the event that claims for recognition conflict with demands for redistribution, and vice versa. Fraser suggests the relative merits can be legitimately assessed through dialogue structured by "participatory parity"—which describes to the material and cultural conditions necessary for every individual to exercise their autonomy as a social equal (36). This concept has strong similarities with the norms of publicity, involving conditions that "permit all (adult) members of society to interact with one another as peers" (ibid.).

In critique, Honneth makes some pertinent points regarding the limitations of Fraser's "bifocal" theory, arguing that it does not encompass issues of political identity and legal injustice—for Honneth, the central conflict dynamic of modern society. He proposes an alternative approach which incorporates social justice and moral philosophy: "the conceptual framework of recognition is of central importance today not because it expresses the objectives of a new type of social movement, but because it has proven to be the appropriate tool for categorically unlocking social experiences of injustice as a whole" (133). Honneth identifies recognition as a normative mainspring for various demands about economic justice, cultural recognition, legal equality, and political representation. He argues that these discriminations pivot on some form of asymmetrical recognition; for example, regarding the denial of respect and denigration of forms of life. Injustices occur when an institutional rule regulating asymmetrical recognition cannot be rationally justified (131).

Notwithstanding their differences, the contributions of both Honneth and Fraser help to advance understanding of the social basis for cross-border identities. The politics of recognition are often framed as a challenge to the state: for example, by either an ethnic claim to self-determination, or for redistribution to remedy even broader political-economic injustices, such as global poverty and environmental degradation. These issues are inherently transboundary. The rising membership of issue-based pressure groups, such as Amnesty International and Greenpeace, testify to the increasing popularity of these concerns—in dramatic contrast to the decline of support for national political parties in many Western democracies. It seems likely that recognition politics can further destabilize state-sponsored conceptions of national citizenship.

In a more pluralistic and information-rich environment, the citizen is exposed to countless "appeals for recognition." Attachment to the national community vies for superiority with many other demands on an individual's loyalty. Giddens argues that the formation of self-identity has become a more reflexive process, because people no longer "rest content with an identity that is simply handed down, inherited, or built on a traditional status. A person's identity has in large part to be discovered, constructed and actively sustained" (Giddens, 1994: 82). It is argued that monolithic institutions like the church and the state are less able to command an individual's exclusive attention. Globalization has expanded the opportunities for the individual to reflect on the beliefs and values of the society in which she is situated, and to take a more decisive hand in shaping her own sense of identity. Again, ICT is a prominent factor in these processes. Lash and Featherstone stress that the transnational politics of recognition rely almost wholly on electronic networking. They argue that: "...global communications are coming increasingly to inhabit the real...Recognition becomes making sense of the information and communitational flows" (Lash and Featherstone, 2001: 17). Likewise, Routledge argues that: "the ability to generate information quickly and deploy it effectively...has become a central component of collective identities of the activists involved, networking forming part of their common repertoire" (Routledge, 2001: 28).

Given the extent of the global digital divide, it should be acknowledged that the "info-rich" has the greatest access to such privileges. It could be argued that cosmopolitan forms of politics are the luxury of the wealthy, "postmaterialist" elite, who are sufficiently secure and well-fed to indulge in abstract philosophical questions about the boundaries of community. Most of humanity faces a constant struggle just to stay alive. These objections merit some sympathy. However, even the most desperate peoples of the world are assuming a more critical stance in relation to national identity. As suggested by Fraser and Honneth, exploited groups can react to globalization by retreating into more cohesive cultural units, and making demands for autonomy or succession from the state. Castells argues that subordinate groups assume a *"resistance identity:* generated by those actors that are in positions/conditions devalued and/or stigmatized by the logic of domination, thus building trenches of resistance and survival on the basis of principles different from, or opposed to, those permeating the institutions of society" (Castells, 1998: 8).

Resistance may be fuelled by heightened cross-cultural awareness. Cosmopolitanism does not always have the positive outcomes extolled by more naïve idealists. Greater exposure to differences in lifestyles and values can often be the cause of widespread anger and antipathy, and inflame reactionary politics. Additional resentment and hostility can be provoked by income disparities and religious differences. A good example of polarized global attitudes was the reaction to 9/11. The notorious pictures of celebrations in the Palestinian streets following the attacks received an indignant and even genuinely bewildered response in the U.S. media (Buck-Morss, 2002: 4). The event yielded the insight that global media can foment international tension. Media messages do not have a uniform interpretation, but are filtered through individual referent points of language, discourse, and culture (Downey and Fenton, 2003: 195). Societies can be connected through global media experiences and yet often separated by vast gulfs of understanding.

This is not to suggest that cosmopolitan movements are the preserve of the "postmaterialist" West, and more parochial forms of politics are borne of anti-Western resentment. Such a crude characterization is belied by the rise of anti-immigration lobbies in Western Europe, or conversely, by the "global egalitarianism" espoused by the Zapatista movement in Mexico. The intention is to highlight general trends, and to stress the fragmentative tendencies of globalization. Indeed, a sense of global commonality is very difficult to cultivate in the context of diverse historical-institutional experiences and intense social heterogeneity. The defensive retreat into resistance identities is an unsurprising reaction to globalization. The revitalization of subordinate cultural and ethnic identities is likely to further weaken the cultural hegemony of nation-states, especially if it stimulates claims for local self-determination. Again, media are central to these processes (Mitra, 2001). As one commentator observes, ICT "make possible a denser, more intense interaction between members of communities who share common cultural characteristics, notably language; and this fact enables us to understand why in recent years we have been witnessing the reemergence of submerged ethnic communities and their nationalisms" (Smith, 1990: 175). Localized identities revived through globalized media are a perfect illustration of the paradoxes inherent to globalization. Some social movements have explicitly sought to reconcile these tendencies, neatly expressed by the slogan: "Think Global, Act Local."

Nation-states have a unique claim on the affiliations of citizens, despite the aforementioned challenges. The "imagined community" of the state represents a powerful amalgam of history, myths, literature, values, and kinsfolk that are matched by few other political institutions. In addition, the legal rights anointed by national citizenship provide substantial foundations for legitimacy. The continued importance of nationhood in people's lives cannot be denied. But what is certainly the case is that people have multiple identities, and that globalizing forces are transforming the relationship between citizen and the state. Nation-states have limited control over the flow and impact of globalized cultural influences, and so increasing pressures have been brought to bear on the primacy of national identity (Jameson, 1991: 322). The outcome of these transformations is ambiguous. Identities are less fixed and less coherent, more pliable than perhaps was the case generations ago. This could be seen as the most subtle and insidious threat to the concept of state sovereignty that has been discussed so far.

To summarize thus far, global governance, the growth of international law and the pluralization of citizen identities have transformed the competencies and public perceptions of the nation-state. The primacy of the state is subject to pincer-like pressures from "above" as well as from "below." It can no longer be assumed to be the sole locus of effective political power. A structure of international authority is emergent that demands the reassessment of "bounded sovereignty" and "international anarchy"—concepts that are implicit in the statist model of the public sphere.

5.3 REVISITING POLITICAL AUTHORITY

The rise of multilayered global governance calls into question the conventional realist notion of the international system as anarchic (Hurd, 1999: 383). International anarchy has generally been described as the absence of international authority. Authority is an endlessly debated concept in the social sciences. In an oft-cited work, Max Weber defined authority as the condition where power is seen as legitimate, where obedience is the norm and transgressions are occasional. Authority may be supplemented by coercion (by force or the threatened use of force), but does not rely on coercion alone to secure compliance. It is distinguished from crude forms of power by the crucial criterion of legitimacy.

Realist theories of anarchy seem to pivot on a perceived lack of legitimacy at the international level. Contrariwise, Hurd suggests that institutions such as the EU are largely regarded as "legitimate" and are therefore an alternative site of authority to the nation-state. Hurd uses the term "legitimacy" to refer to the "normative belief by an actor that a rule or an institution ought to be obeyed" (381). Further, "an actor's belief in the legitimacy of a norm, and thus it's following of that norm, need not correlate to the actor being 'law abiding' or submissive to authority" (ibid.). Rather, an actor's *perception* that a norm is legitimate will be internalized by the actor and provoke behavioral changes. Hurd argues that the persistence of international institutions stems from intersubjective understandings of legitimacy; it cannot be fully understood with exclusive reference to national self-interest or the threat of coercion. The concepts of international authority and global governance help to elucidate the maintenance of global order. The system is built on the basis of institutions and regimes that have some claim to legitimacy.

The "governing institution" of state sovereignty is stable partly because it is widely accepted as legitimate. Norm-internalization by state actors is evident from their habitual compliance with the principle of nonintervention (Liberman, 1993; Russett, 1993). Military deterrence is also an important factor, but not always pertinent. In stable regions, invasion from neighboring states is inconceivable because states define their interests in terms of peaceful coexistence and cooperation. Several states retain nominal armed forces that are insufficient for self-defense in the case of hostile invasion. There is little need—the inviolability of state borders is commonly assumed and observed. Borders are even relatively stable in parts of the world where civil strife is commonplace. For instance, postindependence, the relative stability of the African political map has been quite remarkable—not least because many of the borders were arbitrarily imposed by colonial administrators with little heed to indigenous histories and ethnicities (Jackson and Rosberg, 1982). The survival of these states depends on international recognition of legal sovereign status, rather than their capacity to exercise effective power across their territories. The principle of self-determination prescribes that they are "not allowed to disappear juridically" and "cannot be deprived of sovereignty as a result of war, conquest, partition or colonialism as frequently happened in the past" (Jackson, 1990: 23). The states are propped up by a structure of international authority.

Clearly, expansionist states exist, but they are widely regarded as dangerous aggressors and are outnumbered by states that uphold the status quo. The norm of nonintervention is deeply entrenched (Hurd, 1999: 399). When invasion occurs, it is usually met with an intense hostile reaction from international society—witness the mass outrage over Iraq's invasion of Kuwait. But nonintervention is not entirely sacrosanct. Humanitarian intervention and peacekeeping has become one of the most distinctive features of post–cold war geopolitics. Intervention can be legitimated *in mainstream discourse* if conducted multilaterally and perceived as likely to result in valid humanitarian and security outcomes. The italicized caveat pertains because interference in a state's internal affairs is inherently controversial. Historical prohibitions have been reassessed in exceptional circumstances, but the extent to which the traditional norm of noninterference has been diluted is debatable. The UN has shied away from sanctioning a formal intervention doctrine. As Philpott observes: "... the practice is far from a durable fixture in a new world order. Despite these caveats, however, intervention—widely endorsed and significantly practiced—now seems well within the Security Council's legitimate authority" (Philpott, 1999: 588). Interventions have also been conducted under the aegis of other institutions such as the NATO mission in Kosovo. Again, these are episodic; in Philpott's phrase: "confined to scattered islands of egregious calamity" (589). The norm of nonintervention has been revised somewhat, but the juridical sanctity of the state generally remains resilient. In all interventions, the legitimacy discourse in international society is a key factor in enabling and constraining state actions (Wheeler, 2002). This again underlines the existence of international authority.

National interest is also decisive. I have acknowledged instances above where self-interest can be understood as the prime motivating factor of state behavior. The 2003 coalition attack of Iraq could be seen as egoistic in extremis. Yet even in this case, Hurd's notion of legitimacy has applicability in interpreting the Anglo-American attempt to get a second resolution passed by the Security Council, and subsequent efforts to encourage UN involvement in peacekeeping and "nation-building" initiatives. The global condemnation of the action also revealed that endorsement by the UN was widely regarded as a more valid route to the preservation of international order. In other words, the behavior of the relevant actors indicates that the institution is broadly perceived as legitimate.

If the international system has institutions that are considered legitimate and authoritative, then the concept of anarchy is less relevant to an understanding of global politics. *Less* relevant does not equate to *irrelevant.* Some situations—border conflicts, regional instability—can be most appropriately analyzed with reference to the anarchy thesis. But in the main, contemporary politics is most productively conceived in terms of a global governance model, composed of regimes and institutions that rely on normative influence for compliance. As Hurd argues: "The term *anarchy* seems inappropriate for a system of decentralized authority that actors conform to out of an internal sense of rightness" (Hurd, 1999: 401). The notion of international authority disrupts the traditional internal/external dichotomy on which some of the most fundamental concepts of IR are based. The discipline has been anchored on presumptions about the illimitable, indivisible, and bounded nature of sovereignty. A distinction between the domestic and international realms is still evident, but it is not as clear-cut as portrayed in orthodox accounts. It requires somewhat of an imaginative leap to conceive of denationalized authority. In the words of Alexander Wendt:

> [s]o dominant in contemporary consciousness is the assumption that authority must be centralized that scholars are just beginning to grapple with how decentralized authority might be understood...[T]he question of how to think...about a world after "anarchy" is one of the most important questions facing not only students of international relations but of political theory as well. (Wendt, 1999: 308)

Our intellectual efforts must be reoriented to accommodate different modalities of governance. Sovereignty and territory do not always coincide. There has been a substantial reconfiguration of power and authority, resulting in a transformed political landscape. In the place of the state-centric model is a polyarchic world of multiple sites of authority; in Rosenau's words, a world of "disaggregated complexity" (Rosenau, 2005). The international structure of authority is multilevel and multiactor, with overlapping legislative orbits. The state is a key node in this network, possessing vital functions of decision making and law enforcement. It is not en route to demise, but it has been divested of its predominance as a site of governance.

5.3.1 The Authority Crisis of Global Governance

Thus far, authority has been considered from the perspective of governance elites. Another analytical dimension comes into focus if the legitimacy of international authority is considered from the standpoint of the citizen. There is incongruence between legal frameworks, political communities, and sites of governance, which equates to a democratic disjuncture. This may sound far removed from "bread and butter" issues that are supposed to preoccupy the electorate, but concern about accountability deficits is not confined to the rarefied realms of lofty scholarly discourse. There is a greater sense of awareness of the implications of global governance because citizens routinely encounter the effects of non-state decision making in their daily lives. Global governance is increasingly intrusive, yet not effectively held to popular account. Transnational resistance is on the increase. Notable examples include the "Battle in Seattle" at the WTO meeting, and the "Make Poverty History" rally at the G8 in Edinburgh. Resistance is exhibited in the poorest and the richest societies, in all regions, from authoritarian regimes to mature democracies (which belies the familiar laments from commentators about the levels of "political apathy" and "popular disengagement" in the West). The trend is an expression of the rise of "recognition politics," but also emblematic of a mounting authority crisis in global governance. In the words of one observer: "the societal acceptance of international institutions clearly seems to be in decline…which in turn undermines the effectiveness of these institutions" (Zürn, 2005: 162).

The authority of international institutions is often justified by the principle of "indirect legitimation." In other words, institutions are conferred legitimacy by the citizen in an indirect way, through elected representatives or by appointed officials. The link to the citizen is so tenuous that many feel the democratic elastic has been stretched too far. Moreover, this principle presumes that all member states are democratic, are technically proficient, and there is a rough parity of influence and participation. This is not the case for many underdeveloped states (ibid.). Neither is it true of many international organizations, as exemplified by the inequities of the UN system. Even within the EU, that has a directly elected Parliament, there is widespread dissatisfaction with the perceived remoteness and inaccessibility of the key institutions. In addition, transgovernmental networks and private-authority mechanisms elude effective democratic oversight, despite assuming growing global influence. Cerny summarizes the main issues:

> …[G]lobalization leads to a growing disjunction between the democratic, constitutional and social aspirations of people—which continue to be adapted by and understood through the framework of the territorial nation-state—and the increasingly problematic potential for collective action through state political processes. Certain possibilities for

collective action through multilateral regimes may increase, but these operate at least one remove from democratic accountability...New nodes of private and quasi-public economic power are crystallizing that, in their own partial domains, are in effects more sovereign than the state. (Cerny, 1995: 618)

The pervasiveness of global governance has thereby realigned the relationship between the state and citizen. The relations of responsibility and accountability for global policy are ever more oblique. In the eyes of many, elected representatives seem to have a decreasing ability to shape the political agenda, so it is not clear "where the buck stops." Therefore it is not clear if the state is a consistently effective channel for democratic demands. What is the result of this ambiguity for political life? A widespread sense of disconnect with the decision-making machinery of global governance. A feeling that the value of the vote has been impoverished. And a corresponding growth in the kinds of antiestablishment politics that we shall encounter in the next chapter. International authority thus inheres with a dual tension, deriving from noncompliant states and fractious citizens.

5.4 CONCLUSION

The map of the globe shows clearly demarcated political territories, portraying a tidy order that is easy to assimilate. It is a one-dimensional view of the world that is atheistically and conceptually pleasing. But it is not an accurate depiction of the real state of political topography. The world is more of a hotchpotch of multilayered, overlapping jurisdictions. The effects of globalization on sovereignty and the state have implications for the institutional foundations of public spheres. Habermas' rendition of the bourgeois public sphere was historically and territorially specific, premised on a sovereign and autonomous nation-state. This perspective limits the ability to speculate about the transformation of public spheres outside of the territorial circuits of political power. Habermas has since revised his position, and some of his recent writings have resonance with the key themes of this chapter. Habermas now conceives of a globalized "postnational" realm of multiple publics less restricted by the constraints of material inequality and nationalism (Habermas, 1998b, 2000). He considers that the nation-state once represented "a cogent response to the historical challenge of finding a functional equivalent for the early modern form of social integration that was in the process of disintegrating. Today we are confronting an analogous challenge" (Habermas,

1998a: 398). He argues that the growth of supraterritorial spaces and the consolidation of capitalism as the dominant structure of production is "simply the continuation of a process of which the function of integration performed by the nation-state provided the first major example" (399). Although globalization represents an "unprecedented increase in abstraction we can take our orientation on the precarious path toward postnational societies from the very historical model we are on the point of superceding" (ibid.). He considers the destabilizing of national identities through exposure to ICT, and posits that the growth of international law means that: "[s]tate citizenship and world citizenship form a continuum whose contours, at least, are already becoming visible" (Habermas, 1996: 514).

Similarly, I argue that the exigencies of globalization have propelled the rearticulation of governance. I have illustrated the changing role of the state with reference to the rise of global governance, the growth of international law, and the relative decline of national identities. These processes rupture the notional coincidence of political authority and territory. It is important to reiterate that the effects of globalization are not uniform or homogenous across all societies, nor does globalization equate to the "end of the nation-state." States are juridically robust and potent political actors. One of the lessons of the post-9/11 world is that states also have the capacity to reassert their power in strategically important policy-areas. However, the locus of effective power is context-dependent in the global governance mosaic. In certain issue-areas, private-authority networks may have prime regulatory power; in others, international organizations may have decisive influence; some policy fields may be shaped by the pronouncements of international courts, and so on. Power is dispersed across these different steering mechanisms, forming a structure of international authority. It is a polyarchic world, with minimal coordination between the disparate elements. Global governance is still emergent and its contours are far from fixed. It coexists with older political structures and many of the new patterns of authority are in flux (Rosenau, 1997: 11). It presents a messy and unstable alternative to the idealized coherence of national government.

Transformations in sites of political authority are a structural precondition for the emergence of transnational public spheres. I posit that this precondition has been met. Global governance provides an alternative political-institutional framework to the state for public dialogue and mobilization. As a collary, I propose that it has eroded the critical function of domestic public spheres—to exert meaningful influence over decision making. Transnational oppositional politics is on the rise, which suggests a popular authority crisis in prevailing forms of governance.

However, there are two main problems with the political backdrop to emergent publics. First, the decision-making process is inchoate. The relationships between institutions are sometimes marked by tension and conflict, which produces confused policy outcomes. It is unclear where power lies in such amorphous governance arrangements. Second, global governance does not have a demos, or well-defined political community. Therefore, is also unclear who decision-makers should be answerable to, or indeed if public spheres would be functional outside of a shared political culture. In short, the relationships of account-ability are abstruse, and the notion of transnational publics problem-atical. Let us then turn to investigate the social underpinnings of extraterritorial publicity.

CHAPTER 6

Global Civil Society: Transnational Networks of Mutual Affinity

The seeds of political emancipation may be rooted in global civil society. Transnational social movements are now a permanent feature of the world political landscape (della Porta et al., 1999: 206). Countless social movements habitually network across state borders to exchange information, to debate political issues, to develop strategies and policies, and to solicit transnational support (Cohen and Rai, 2000). Networking seems to typify the spirit of the "information age," but it is not a twenty-first century invention. Historical precursors for transnational activism are vividly illustrated by Keck and Sikkink's analyses of the Abolition Movement and the International Suffrage Movement (1998). In addition, Aravamudan's (1999) analysis of colonial-era literature, Gilroy's (1993) discussion of black vernacular cultures and Linebaugh and Rediker's (2000) study on the multiethnic Atlantic working class expose the oft-neglected deliberative history of other subordinate groups. However, transborder networking has since acquired more visibility and greater political prominence. This transformation is partly a by-product of the increased accessibility and global scope of ICT. Most civil society organizations will have some kind of Internet presence, which opens up prospects of making contact with sympathetic individuals and groups from far afield (Castells, 1997: Chapters 2 and 3). Transnational networking is also partly a response to the evolving architecture of political authority. Iconic junctures of the global governance process are routine targets for vibrant demonstrations by international coalitions of activists. Meetings of the WTO and the G8 are rarely unattended by mass demonstrations on the outside. Caroline Thomas suggests that increased interconnections are best conceptualized as "the process whereby power is located in global social formations and expressed through global networks rather than through territorially-based states" (Thomas, 1997: 6). This techno-political context engenders new forms of civic engagement whereby the norms of publicity may be progressively instantiated.

An institutional prerequisite for emergent transnational public spheres is networks of political activists in diverse locales, interlinked by regular ICT usage. Transnational networks can be assessed according to the criteria of *mutual affinity, norms of publicity,* and *political efficacy.* The first criterion—*mutual affinity*—refers to a minimal sense of commonality and trust between the participants. Mutual affinity is a necessary condition for the reciprocal recognition of the moral-political validity of deliberative norms. These cosmopolitan ideals are exceptionally difficult to achieve amongst transnational networks that are geographically dispersed across several political territories. The criterion of *norms of publicity* assesses the extent to which dialogue aspires to the ideals of public sphere discourse. The anonymous methods of communication that typify Internet discourse can be antithetical to the requirements of normatively structured deliberation (Calhoun, 2003). It is interesting that despite such difficulties, global civil society appears to be in rude health when compared to ailing mainstream political parties in established democracies, both in terms of active members and in levels of public trust and confidence. The deliberative quality of these networks plainly bears closer scrutiny. It is also valuable to assess the impact of network dialogue on the mainstream political agenda and global governance. As Bohman argues, the critical function of the public sphere ultimately rests on the connection between public opinion and political authority. The criterion of *political efficacy* refers to the strength of this relationship. This chapter evaluates the extent to which all three criteria are met with reference to the activities of numerous networked social movements.

The discussion is structured in four parts. First, global civil society theory is critiqued, and the distinction between the concepts of public sphere and civil society is assessed. Second, the "virtual communities" literature is reviewed, and I propose "networks" as a more satisfactory term for an exploratory investigation into mediated social relations. Third, there is a more detailed explanation of the analytical criteria outlined above to contextualize the subsequent case studies. Fourth, three main subject areas provide microcosmic examples of how embryonic public spheres may emerge through transnational coalitions of networked citizenry: the international women's movement, the Zapatistas, and Greenpeace. Each subject area exemplifies a distinct social-political basis from which the participants can derive a sense of mutual affinity. In the case of the women's movement, the basis is gendered experience, in the case of the Zapatistas, it is anti-neoliberal rhetoric, and in the case of Greenpeace, it is the ecosystem. I conclude the chapter by arguing that transnational networks of mutual affinity provide a structural precondition for the emergence of transnational public spheres.

6.1 ON THE RELATIVE MERITS OF THE PUBLIC SPHERE APPROACH

If globalization has rejuvenated civil society, perhaps public spheres could also be revitalized. The possible juxtaposition has not escaped the attention of Habermas. As previously discussed, he has recently hailed the resurgence of civil society and mused over prospects for critical publicity in this promising climate (Habermas, 1996: 330). The similar meaning of the "public sphere" and "civil society" prompts a cautionary note—although they are inextricably linked, they should not be understood as interchangeable terms. Before proceeding, it is worth restating the distinction and emphasizing the distinct nuances between the concepts. This discussion has been foreshadowed in section 2.3; but I want to elaborate further on the specific advantages of public sphere theory in order to frame the successive case studies in the broader argument.

Anheier et al. define "global civil society" as "a supranational sphere of social and political participation," distinct from the practices of governance and economy, but existing "above and beyond national, regional and local societies" (Anheier et al., 2001: 4). Likewise, Scholte suggests that: "we can take 'civil society' to refer to those activities by voluntary associations to shape policies, norms, and/or deeper social structures. Civil society is therefore distinct from both official and commercial circles" (Scholte, 2000: 277). The conceptual similarities here with public sphere theory are evident. Both approaches focus on the groups that mediate between society and the state, providing an essential "buffer zone" between citizens and sovereign power. But the "public sphere" is laden with a more specific meaning: it describes the *process whereby political authority can be subjected to normative critique*. Public sphere theory invites inquiry into effective forms of social organization within civil society for reflexive dialogue. It requires the full participation of all affected actors and the adjudication of disputes through rational argumentation. This is not to suggest that these concerns will not interest global civil society theorists—rather to reiterate that this theme is integral and inherent to public sphere theory.

Civil society theory presumes communication between actors, but does not similarly rest on a theoretical ideal of dialogue and explicit norms of inclusion and participation. Instead, it tends to focus on the extent to which the policymaking process is shaped and decision-makers are influenced by civil society. In the words of Anheier et al., the key task for an agent of global civil society is "about increasing the responsiveness of political institutions...the need to influence and put pressure on global institutions in order to reclaim control over local political space" (Anheier et al., 2001: 11; also see Falk, 1998). These issues also have relevance for public sphere theory. Nevertheless, a public sphere perspective dictates that it is essential to critically reflect on the conditions of communication within civil society; otherwise discursive power relations become progressively entrenched. For example, consider the tendency in the global civil society literature to treat social movements as unproblematic agents of emancipation. As Amoore and Langley observe, "implicit in much of the academic advocacy of [global civil society] is the belief that by acting as a progressive force for 'good,' [global civil society] provides the key to resistance in the contemporary world order" (Amoore and Langley, 2004: 98). Drainville (1998) also disparages the "fetishism" of global civil society, whereby social movements en masse are a repository for hopes of counterhegemonic struggle.

There is a danger in this literature that civil society is being portrayed as the unambiguous embodiment of liberal norms, such as human rights and democracy (Keane, 2001: 57). This approach shows a disregard for the implications of significant variations in internal deliberative and decision-making procedures, which can fall considerably short of public sphere ideals (e.g., Colás, 2002; Warkentin, 2001). The uncritical approach also effectively discounts the differing political aims of social movements. Many are extremely repressive and reactionary. As an extreme example, the neo-Nazis and the Taliban have been very active online (Chroust, 2000). It is also the case that racism, sexism, and other prejudicial attitudes are present in other movements widely regarded in mainstream discourse as "legitimate." These unpalatable realities are disguised when civil society is solely or largely characterized as a source of transformative agency. As Pasha and Blaney argue, the "notions of civility that are increasingly attached to civil society, while enabling a certain form of civil life, also contribute to a narrowing of the political agenda and the exclusion of certain actors and voices" (Pasha and Blaney, 2001: 423). The agenda is further distorted when global civil society is represented as a cohesive agent, as if a common consensus existed between social movements over global issues. This is profoundly at odds with the facts. Global civil society is as richly diverse as the humans of which it is constitutive. Yet the civil society organizations that achieve mainstream legitimacy tend to be "established after the image of the civilised (European) male individual" (Keane, 1998: 21). Marginalized voices of dissent are either excluded from mainstream dialogue, or else are in danger of being co-opted into hegemonic discourses and losing authenticity. It is important to remember that civil society is constituted by the politics of power and control, as well as the politics of emancipation and resistance.

The choice of approach—global civil society or public sphere theory—is dependent on the questions that one is interested in asking. Civil society theory neither precludes analysis of discursive relations of domination and inequality nor is it antithetical to an examination of the relationship between ICT and politics. But such analysis can be problematical because civil society theory is anchored on issues that differ from those that underpin this inquiry. Public sphere theory is more suitably designed for my interests in equitable participation and the sociopolitical effects of ICT.

There are peculiar difficulties with applying a public sphere framework to civil society groups online. Electronically mediated communication has specific qualities that some perceive as hostile to normative political deliberation. The controversy revolves around whether meaningful social bonds can be forged through virtual interaction. I turn to consider how this debate has evolved in the "virtual communities" literature.

6.2 Theorizing Virtual Communities

"Community" is one of the quintessential "contested concepts" of social science. Its definition and meaning have consistently been subjects of intense debate amongst sociologists and anthropologists. In a seminal work, Tonnies (1957) conceived of the transition from preindustrial to industrial society through a dual typology of human interaction. He posited that earlier times were characterized by gemeinschaft: physically copresent groups of people bound by feelings of kinship and solidarity. The process of industrialization instigated a transformation toward gesellschaft: a rise of impersonal relationships characterized by rationality, instrumentality, and calculability. For Tonnies, this progression signaled the degradation of community. The thesis has been considerably influential; indeed, Habermas' subsequent distinction between "system" and "lifeworld" can be understood in parallel. Consequently, physical location has conventionally been emphasized as an essential component of a "thick" community identity, or gemeinschaft. In comparison, online networking is oft-portrayed as the epitome of gesellschaft.

However, simplistic caricatures are unfair because there are real differences between forms of online networking. Virnoche and Marx (1997) distinguish between "community networks" (that reflect and reinforce preexisting ties), "virtual extensions" (that increase opportunities for interaction among preexisting communities), and "virtual communities" (spatially different actors, where ICT is the primary or sole means of communication). The latter represents the strongest challenge to the notion of transnational publics, as virtual communities are not underpinned by common citizenship, shared locality, or physical interaction. They are abstracted from factors of proximity and are based on information exchange, deliberation, shared norms, and intersubjective understandings. It seems to be this type of networking that critics intend to target when they use the term gesellschaft (e.g., Calhoun, 2003: 243–244).

For example, Foster (1997) identifies the difference between gemeinschaft and gesellschaft as a regular flow of "we-relevant" information that provides a sense of collective identity, which he argues is present only in the former. He explores the role that the relationship between the individual and society has in the establishment and maintenance of a community, and sees a reflexive connection between self-identity and group identity. Without a conception of the self, he argues, community is unrealizable (and conversely the community has considerable influence on the development of the self). Foster is therefore skeptical about the potentialities of virtual community because he observes that Internet users tend to be interested in communicating details about themselves, rather than contributing to the flow of "we-relevant" information that forms part of the collective effort of sustaining community relations.

Likewise, Johnston and Laxter maintain that a virtual network of geographically dispersed individuals is a poor substitute for "face-to-face-social movement connections" (Johnston and Laxter, 2003: 73). Sidney Tarrow also expresses doubt that virtual networks can deliver the same "crystallization of mutual trust and collective identity" that characterizes the interpersonal ties of localized social movements (Tarrow, 1998: 14). This line of argument is indicative of an unease that demonstrates "the continued power of the territorial narrative and the appeal of 'real' places" (Axford, 2001: 18). The suspicion is that online networking is not "meaningful" enough to build a sense of communal identity across state borders. The experience of being at home alone with the Internet is portrayed as "asocial, instrumental and narcissistic" (ibid.). "Thick" communities are depicted as more "authentic," the relationships within as more genuine, and loyalties as more solid than anything that can be created or sustained online. The subtext is that "thick" forms of community provide the only legitimate basis for political activity.

No doubt, identities associated with physical copresence can be strongly embedded (McAdam et al., 2000; Melucci, 1996). But it does not necessarily follow that location is a prerequisite for fostering a community spirit. Norms, shared experiences, and communicative interaction are equally important for social bonding. In his seminal work on American communities and social change, Bender critiqued sociological analysis that privileges shared territory at the expense of culture:

> The identification of community with locality and communal experiences with rather casual associations has quietly redefined community in a way that puts it at odds with its historical and popular meaning...drain[ing] the concept of the very qualities that give the notion of community its cultural, as opposed to merely organizational, significance. (Bender, 1978: 10)

Indeed, the relevance of location to community cohesion has been problematized in social science long before the advent of the Internet (e.g., Scherer, 1972). The topic has acquired increased relevance because ICTs expand the potential for community formation amongst those with a commonality of interests. It poses a fundamental challenge to conventional understandings of sociability, temporality, and spatiality. It is useful here to recall Harvey's conception of "time-space compression," which refers to the ways in which popular intersubjective understandings of the world have undergone radical revision owing to the dismantling of temporal-spatial barriers. For example, notions like the "global village" have gained popular currency. These globalizing discourses implicitly emphasize humanity's interconnectivity, and can influence the way in which people perceive their relationships to distant others. Perhaps the notion of delocalized community may be more readily accepted in conjunction with such concepts.

Virtual networks are abundant, ranging from fan clubs to cultural diasporas, where people habitually converse, form friendships, and even fall in love with one another. These everyday experiences cannot be readily dismissed as hollow products of gesellschaft. Axford notes that ICT can "sustain the activities of a tranche of INGOs (International Non-Governmental Organizations) and social movements…and provide a degree of information and support for a host of people ill served by the public services in the 'real' civic spaces where they live out their lives" (Axford, 2001: 18). In short, new media does not necessarily promote anomie. Instead, virtual space can facilitate social bonding for those who experience a degradation of community in their physical living spaces. In the words of one enthusiast, virtual communities are "incontrovertibly social spaces in which people still meet face-to-face, but under new definitions of both 'meet' and 'face'…[V]irtual communities [are] passage points for collections of common beliefs and practices that unite people who are physically separated" (Stone, 1991: 85).

But skeptics of virtual communities question whether delocalized, text-based discussion can ever be substantial enough to be under-stood as meaningful (e.g., Jones, 1995; Miller, 1996; Negroponte, 1995; Shields, 1996). We have already encountered Bohman's misgiv-ings about "anonymous" forms of Internet communication, where: "…we do not know who is actually speaking; we also do not know whom we expect to respond, if they will respond or if the response will be sustained" (Bohman, 2004: 138). It is true that many Internet encounters are arbitrary and transient, and thus of little deliberative value (Noveck, 2000). There are often problems in verifying the iden-tities of authors. Users can post messages in forums without revealing their true names, or by assuming false personas. Bob, the regular middle-aged trucker from Chicago can adopt the guise of Barbara, the young and glamorous model from New York. Optimistically, anonymous interaction can be interpreted as an opportunity for lib-eration from our bodily selves, with all the attendant prejudices that society ascribes to our gender, ethnicity, class, age, and so on. Remove these constraints, one might argue, and competing claims can be more fairly adjudicated on rational merit. However, a cursory glance at online chatrooms reveals that many take advantage of the cloak of anonymity to indulge in abusive behavior. Freed from the restraints of accountability, people can use cyberspace to insult and libel others without fear of redress.

Face-to-face interaction has been portrayed by some as more "authentic" by comparison. For example, Poster contends that identity is defined by the physical body and by contact with others, and that the Internet is little more than a phony simulacrum of "real life": "Without embodied copresence, the charisma and status of individuals have no force...The technology of the Internet shouldn't be viewed as a new form of public sphere" (Poster, 1995b). Poster's notion that public spheres are dependent on the proximity of interlocutors must be challenged. As argued in chapter 4, disembodied discourse has always been a feature of the public sphere. For example, in the bourgeois public, the medium of print was necessary for participants to communicate over long distances. The criticisms regarding the characteristics of Internet discourse are more profound. There are genuine problems in ascertaining authenticity in electronic text. How can one establish the intended meaning of a message when one is unsure of the author, nor has access to the extra information that verbal and nonverbal communication would convey? It is difficult to engender an environment for the public use of reason when there are few mechanisms for building trust.

In an instructive study, Watson (1997) argues that whilst online interaction has radically different qualities from face-to-face interaction, it is not necessarily an inferior form of communication. He examines peculiar cultural features of Internet interaction, such as so-called emoticons, which denote emotional responses and reactions via depictive text icons. Watson sees these as methods of mutual understanding that are indicative of evolving community. They suggest that participants are interested in increasing the authenticity and sincerity of anonymous interaction. Watson emphasizes that many virtual communities are linked by similar interests, which aids social bonding. For its members, the virtual community can be regarded as an experience just as "real" as the physical community.

Disputes about the significance of virtual communities will no doubt continue for the foreseeable future. But as virtual networks become more pervasive and established, it also seems possible to discern a greater respect in academia for their contribution to modern political life. In the words of one observer: "current social and organizational ties seem to be weak in their form and stability, whereas the sphere of mediated communication, especially through the Net, appears to be widening and strengthening" (Sassi, 2001: 100). Nonetheless, because of wide variances in form and content, it seems wise to avoid the term "community" to refer to sites of virtual discourse collectively. I do not reject the notion of virtual community, but a more neutral noun is needed for an exploratory investigation into emergent transnational public spheres. "Community" can be interpreted as laying a greater claim on close social bonds than a nascent public requires. "Networks" is altogether preferable. A public *does* need to have a sense of mutual affinity, because they will not engage in rational-critical discourse without a normative regard for the other. However, in a transnational, electronically mediated environment, mutual affinity is likely to be minimal.

6.3   On the Criteria for Transnational Networks of Mutual Affinity

It is worth recalling the main elements of public spheres before proceeding. I have proposed a definition of a public sphere that was specifically designed to permit generalizations across different historical experiences and cultural environments. It is my attempt to tackle a problem central to transnational public spheres theory: how publicity may be conceived in the context of cultural heterogeneity. Bohman has considered this at length and suggested that social acts are "public" only if they meet the following basic requirements: "[t]hey not only must be directed to an indefinite audience, but also offered with some expectation of a response, especially with regard to their intelligibility and justifiability to others" (Bohman, 1998: 207). Further, "public actions constitute a common and open 'space' for interaction with indefinite others…which can be broader or narrower in comparison with others in terms of topics, available social roles, forms of expression, and so on" (ibid.).

Let us consider these requirements in more detail, particularly in regard to the "publicness" of an utterance. It can be difficult to distinguish between public and private discourse in a fragmented virtual realm. The difference is partially situated in message-content. Private communication is directed at a specific and demarcated audience. Private interlocutors share specific understandings about how truth-claims can be made, and who is authorized to make them. Different standards operate in public address, which presupposes a potentially unrestricted realm of communication (Bohman, 1997: 183). The public do not share assumptions about truth-claims; therefore if truth-claims are submitted, they need to be subjected to rational argumentation before securing wider agreement. Deliberative norms require that differing reasons and judgments are evaluated against one another through a free interchange of views between speaker and audience. One is expected to enter into reasoned discourse and to account for one's opinions and actions in an intelligible manner. In addition, the norms of publicity invalidate arbitrary claims about who should be excluded from dialogue. This ensures the maximal inclusiveness of public spheres, and debars discourse that is reactionary, bigoted, and prejudiced.

Public sphere theory thus pivots on issues of participation. At the highest level of abstraction, all participants should be of equal status as deliberative actors. In reality, this occurs rarely, if ever. Public sphere ideals demand universal access, but actual publics assume specific forms that are often inscribed with power disparities. Factors such as educational background and social inequality make truly egalitarian participation impossible (behold the exclusions in the constitution of the bourgeois public). Nonetheless, rational-critical dialogue should at least be framed by an inclusive perspective and accompanied by attempts to integrate the diverse concerns of all affected actors, including the vulnerable and marginalized.

I have proposed a multiple spheres model to accommodate a variety of potential manifestations of specific forms of publics. The transnational realm is highly pluralistic, as it lacks the coherence of national public spheres in terms of moral and political community. Social fragmentation is the unavoidable consequence. The public use of reason in these circumstances can be understood as "a form of publicity that capaciously allows for multiple forms of publicity within the contours of its multilayered and differentiated social space" (Bohman, 1998: 210). But debate in these spheres can broadly be considered as "public" if it is not arbitrarily restricted to an exclusive audience, and if it is inclusive in orientation. In short, discourse should be addressed to an audience of infinite strangers.

Habermas describes a public sphere as "private persons come together as a public" (Habermas, 1999: 27). The social conditions underpinning delocalized publics are problematic. Bohman outlines the main difficulties when he asserts that cultural differentiation and the anonymity of computer-mediated communication render transnational audiences as little more than aggregate publics. But perhaps this conclusion is too pessimistic. As discussed in section 5.2.3, the rise of recognition politics illustrates how the character of political activism has changed in recent years. People routinely form political groupings based on common claims to recognition that transcend the nation-state. This indicates an immanent potential for the expansion of normative discourse that invites closer examination. The question is whether it is possible for participants to collectively identify with a social imaginary of a "public" that is multicultural, denationalized, and based on virtual discourse.

I have found Dahlgren's work useful in conceptualizing these issues. His model of civic culture involves six elements: values, affinity, knowledge, practices, identities, and discussion (Dahlgren, 2002). He posits these as an analytic framework for future empirical study. Dahlgren is also influenced by Habermasian theory and his themes intertwine with mine, although our perspectives are far from identical. For example, my approach can be contrasted by the central claim to the structural prerequisite of global governance, and by the heavy reference to *STPS* and the IR literature. Most of Dahlgren's parameters of civic culture can be roughly subsumed under the broader criterion of the norms of publicity, but the category of "affinity" is a productive point of engagement between our conceptual frameworks. He describes *affinity* in terms of a limited "sense of commonality and trust," specifically to evade the implication of "[c]ommunity of the more compelling kind, with strong affect..." (17). Affinity describes:

> ...a minimal sense among citizens in heterogeneous late modern societies that they belong to the same social and political entities, despite all other differences, and have to deal with each other to make it work, whether at the level of the neighbourhood, nation state or the global arena. (ibid.)

It is a neat way of encapsulating the tension between integration and fragmentation that characterizes global social relations. Dahlgren proposes a "modest, minimal threshold" that effectively counters Bohman's skepticism of virtual publics. By "deliberating avoiding a communitarian argument for a foundation for democratic society," the approach also averts controversies surrounding "virtual community," but is compliant with "virtual networks" (ibid.). For Dahlgren, *affinity* describes a reciprocal recognition of the need for interaction between citizens, which leads to the formation of cooperative networks. In other words, a sense of mutual affinity engenders conditions for new social imaginaries of the "public" to emerge. It can therefore be surmised that the emergence of transnational public spheres is dependent on a sense of affinity between the interlocutors. I call this the criterion of *mutual affinity*.

The strength of these social bonds can be partly divined through the deliberative quality of interaction. Critical publicity should be characterized by intelligibility, accountability, and inclusiveness, and be motivated by a desire to rationalize sovereign domination (these are the criteria of the *norms of publicity*). Transnational networks of mutual affinity must exhibit these norms in discourse to provide the social precondition of a transnational public sphere. There are some examples of case study analyses of mediated social movements, where deliberative quality is assessed according to Habermas' theory of communicative rationality (e.g., Dahlberg, 2001; O'Donnell, 2001). The research is interesting, but it is based on the premise that that transnational public spheres are actually existing institutions. I am not concerned here with Habermas' quasi-transcendental "linguistic turn." My ambitions are more modest, as it is not evident that the structural preconditions exist in the present for the emergence of transnational critical publicity. My intentions here are to assess the general indications of the emergence of deliberative norms through a survey of transnational social movement activity.

I am also interested in evaluating the political impact of dialogue in civil society. Emergent public spheres should not be purely a location for critique but also a source of societal transformation. In Bohman's words, public spheres can "...be dynamic enough to reshape the framework of existing political institutions to require acknowledgement of the rights of members of the universal community outside the boundaries of its territory and its membership" (Bohman, 1997: 187). This relationship to governing institutions is the primary characteristic of a public sphere. It is what infuses the public with critical value and political power. For example, feminist counterpublics were crucial in the push toward universal suffrage. Thus, it should be possible to identify transformations in the global governance framework that can be partly attributed to public dialogue. It may be exhibited in a variety of ways. Perhaps decision-makers directly consult with social movements, a public campaign reaches a successful conclusion, or so much publicity is secured for a counterhegemonic discourse that the political agenda is altered as a result (Warf and Grimes, 1997). This function of the public can be summarized as the criterion of *political efficacy*.

The three transnational social movements examined in this chapter each have a claim for recognition on the basis of a different interest or identity, thus constituting a rigorous test of the notion of mutual affinity. The women's movement obviously appeals to a shared experience of gendered oppression. The Zapatista rebellion was a localized dispute that has garnered international support under the banner of anti-neoliberalism. Discourse within this movement has been frequently addressed to exploited classes worldwide. Greenpeace argue that ecological degradation is a universal concern, and appeals to a sense of common humanity. It is worth noting that transnational activists may be aligned with several social movements, and so communities of mutual affinity might not be discrete but rather overlapping. For example, one could theoretically be a supporter of the Zapatista movement, a feminist, and an environmental activist. These complex identities are accommodated in the multiple spheres model.

Each case study is also of interest in other regards. First, the women's movement has a tradition of cultivating cross-border solidarities, and has been a source of radical and transformative discourse (Landes, 1998; Ryan, 1992). Women have fought to bring subjects traditionally regarded as "private" into the public realm, relating to sex, marriage, the family, and the household (Landes, 1988). The historical contribution of the women's movement to the development of the public sphere has already been reviewed—below, the narrative continues. Second, the Zapatista rebellion has become a "primary reference point" for the study of networking by transnational social movements, as "the most striking thing about the sequence of events set in motion on January 1, 1994 has been the speed with which news of the struggle circulated and the rapidity of the mobilization of support which resulted" (Cleaver, 1999: 2). The Zapatistas displayed considerable ingenuity in using ICT to build international solidarities with other activists and grassroots organizations, and became a "prototype" for other networked social movements (Arquilla and Ronfeldt, 1996: 73). Third, the international environmental movement is growing fast and is increasingly active, and has been a subject of keen scholarly interest (e.g., Mancusi-Materi, 1999; Paterson, 2001; Pickerill, 2003; Wapner, 2002). Greenpeace is perhaps the best-known environmental organization, and maintains one of the most extensive and well-maintained NGO Web sites in the world. It has made pioneering use of ICT to facilitate dialogue between its activists through its "Cyberactivist Center." The center has hosted insightful discussions concerning the benefits and drawbacks of online networking. Each of these case studies will be examined in turn, and evaluated against the criteria of *mutual affinity*, the *norms of publicity*, and *political efficacy* in the concluding section.

6.4 The International Women's Movement

The women's movement has a proud history of submitting androcentric notions of politics and society to robust challenge (McLaughlin, 1993). Transnational networking stretches back to the international campaign for women's suffrage from the late nineteenth century onward (Keck and Sikkink, 1998: 51–58). As communication technology has evolved, so has the capacity for international activism. As Youngs observes,

> One of the major successes of women's global networking in the Internet age has in fact been to connect effectively the different kinds of communication technology available at different times and in different places so that the influence of the Internet reaches far beyond the limits of its actual connectivity. (Youngs, 2002: 26)

This section provides a broad overview of how women's groups have used ICT as tools of empowerment: to transcend sexual, religious, ethnic, and spatial barriers, to combat exclusions and to break down walls of silence. Youngs contends that to "the extent to which the virtual realm of the Internet has offered opportunities for some disruption of the masculine domination of international political space(s), it is both actually and potentially revolutionary for women and the larger political scene" (25). ICT enable new forms of collective politics by providing an infrastructure for transnational networking that was previously absent, or by superseding an older infrastructure that was inferior in terms of speed, scope, and interactivity. "Real" place-based politics are often closely related to the virtual politics of female activism, as demonstrated by the multiple case studies below. The first part of this section outlines a series of notable examples of networking, political mobilization, and online campaigning by NGOs from across the world. The second part considers the engagement of women with global governance institutions, by examining the activities surrounding the UN Beijing conference. It analyzes the legacy of the conference in promoting wider ICT usage throughout the movement.

6.4.1   Transnational Networking for Female Empowerment

The Internet was a heavily male-dominated medium when first introduced to the public. The feminist community initiated a campaign for improved female access, which was met with particular success in the United States. In 1995, only 15 percent of American Internet users were women, yet this had increased to 50 percent just five years later (Richards and Schnall, 2003). The United States hosted some of the earliest examples of feminist Web sites, such as those run by the National Organization for Women (NOW), Feminist Majority Foundation and Feminist.com. The latter is still one of the foremost Internet resources for the women's movement. It has played a significant part in promoting women's access by offering a free Web presence to groups that are not yet online. Its primary goal is to facilitate networking for women, or in the words of the mission statement, to "become a place where women can meet, exchange ideas, get information, build a business, become active, participate in government and empower themselves and the world around them—a woman's cyber community if you will" (Schnall). The campaign has had unintentional benefits in raising women's wider political awareness. For example, Richards and Schnall (2003) have reflected on how responses to Feminist.com have revealed an undercurrent of ignorance and despair amongst American women suffering job discrimination. Women had been flooding the site with complaints that federal legislation did not apply to companies with less than 50 employees—in fact, the legislation *was* applicable, and the women had been misinformed by employers. It appeared as if many complainants had experienced their situation in isolation, and until they had consulted the site, they had not realized that such discrimination was endemic. By sharing their experiences of injustice, the women corrected their misconceptions and forged political solidarities. The site highlighted the need for alternative job discrimination watchdogs to be established and run by women's groups in order to pressurize employers to meet their legal responsibilities. Richards and Schnall describe other indirect benefits of an Internet presence:

> People searching for "custody" or "unequal pay" or even "female road-sters" can be virtually introduced to feminist resources without having realized that feminism is what they needed, after all. *They get the chance to grasp their connection to feminism without first having to confront it and overcome their biases against it.* The process itself demystifies feminism. (Richards and Schnall, 2003, original emphasis)

This is interesting because theorists such as Bohman often criticize the randomized, ad hoc connections fostered by the Internet as antithetical to meaningful interaction. In contrast, Richards and Schnall attest to the advantages of being exposed to discourses that one may not otherwise encounter. But is it possible for computer-mediated communication to foster feelings of mutual affinity between remote persons? In a cyber-ethnographic study of feminist online communities, Ward (1999) asked the participants of feminist site Cybergrrl about their experiences. The responses are worth reproducing here:

> The Village is really a place for us all to share thoughts and ideas, meet people and really connect.
>
> I find the Cybergrrl site to be a very positive place for women to explore the internet and participate in the creation of a community.

I love the way we can all meet up here and there are so many people from such different backgrounds...it's so cool!

It's so weird it's like we're all just sitting at our computers and we have created this world. It's almost spiritual.

I feel like it's a community, sort of. I know some people by name and suspect they know me, too, although I've never directly talked to them...I know that some people read what I write and that gives me immense satisfaction. There are people around to help if there are problems. There are also black sheep around, and that's what rounds the picture. I find it difficult to keep closer relationships going in cybergrrl, but I have that problem in RL [real life] too. What else does a community need? (Ward, 1999: 9)

As these extracts illustrate, despite the anonymity of Internet communication, people can experience feelings of affinity online. It also demonstrates that social diversity does not necessarily frustrate a sense of commonality (Kennedy, 2000). The members expressly appreciate the participation of a range of people from "different backgrounds." Also, feelings of affinity do not seem to be negated by delocalized interaction, as vividly demonstrated by the fourth quote. In fact, the ability to communicate across vast distances appears to add to the vibrancy and dynamism of the community. The final quote is interesting in that it draws direct analogies between the physical and virtual worlds, suggesting that problems experienced online can also be replicated through face-to-face interaction (Nip, 2004).

Of course, undeniable differences result from the transience of virtual networks compared to the relative stability of physical communities. Members of networks like Cybergrrl tend to be ephemeral and self-interested, in that they tend to become involved when they experience problems and need advice and support. Although it could be said that such an attitude contradicts the traditional "public minded" spirit of community, it is important not to "overidealize" the experience of physical communities at the expense of virtual networks (Watson, 1997). People are simultaneously members of a number of different communities in their everyday lives; for instance, one may be involved in a local community group, a church, and a sporting team. Across time, people frequently choose to vary their level of participation in different groups according to their personal needs and wants. This trend may be reflected in the virtual world as well. Indeed, the participants of Cybergrrl and other similar sites seem to regard variable participation by fellow users as quite acceptable. As Ward observes: "It seems that people enjoy the diversity they experience within their online communities and they overtly admit that they are being thoroughly instrumental in their choice of community" (Ward, 1999: 12). Indeed, the personal testimonials above pay tribute to the great benefit that members derive from their mediated relationship with others. It appears that despite the spasmodic nature of their participation, members of Cybergrrl derive feelings of affinity from the basis of shared identities and interests.

This general principle can be said to be true of many similar virtual networks (Smith and Kollock, 1999). But Harcourt et al. argue that women's networks have distinct characteristics because of shared gendered experiences. While acknowledging cultural differences, they maintain that it is increasingly possible to identify a "common purpose" amongst women activists embroiled in political conflicts of home and body:

> A common language has emerged from long years of local, national and international struggles that renders their goals more legitimate and transparent, and adds to the strength of their appeals. From the local experience of place-based politics and the exchange and mutual support that occurs in networking, strategies for their struggles are crafted and activated. (Harcourt et al., 2002: 44)

Some of the most interesting examples of this type of interaction are located in the underdeveloped world, where women's groups have to contend with chronic underprovision of female education and poor ICT infrastructure. The ways in which ICT has been applied varies considerably owing to the specific project goals, the location of the group, the skills of members, and the nature of the audience with which they seek engagement.

For example, the Women on the Net (WoN) project was established to deal with the differential needs of women in high-income and low-income countries. It was a joint collaboration between several women's groups, the Society for International Development, and UNESCO. The project was designed to contribute to a new Internet culture from a gender perspective. Its main aims were threefold: to encourage and empower women in the global South and marginalized groups in the North to use the Internet as a political tool, to formulate an agenda for the transnational women's movement regarding telecommunication policies, and to create a community and support base for women to encourage more effectual use of the Internet. Harcourt was a leading member of the project and reported that the virtual network had cultivated "an intimacy which the solitary act of typing into a keyboard in front of a screen belies but which the ethos of WoN embraces" (Harcourt, 2000: 54).

WoN have held small training workshops in various locations in the South to introduce local women to the Internet, most of whom had never seen a computer before. Issues such as domestic violence and sex education are explored in the workshops, often in the midst of cultures where such subjects are normally relegated to the private sphere and considered taboo (ibid.). These social and educative programs often act as a "safe" public space for women to reflect on gendered power relations, freed from the confines of house and family. The conversations can instigate ripples of social change that are eventually felt widely in the whole local community (Knouse and Webb, 2001). Harcourt asserts that personal and social transformations can be reinforced by Internet interaction. She cites WoN colleague, Peggy Antrobus, the head of Development Alternatives with Women for a New Era (DAWN), thus:

> Internet most of all…has empowered us, by giving us the information, the analysis, the sense of solidarity, the experience of shared achievements, the encouragement and moral support that comes from being part of a network, a movement with common goals and visions. (Harcourt, 1999: 13; also see Antrobus, 2004)

Several women's groups in the underdeveloped world have managed to assume an online presence even in the absence of a regular Internet connection. They have deployed different media strategies according to their needs and resources, often demonstrating considerable creativity and inventiveness to overcome numerous obstacles (Mbambo, 1999; Youngs, 2000). This is exemplified by the case of Sakshi, an Indian women's group with very limited Internet access. They appealed to members of the Association of Progressive Communications (APC) Women's Networking Support Program for help with their lobbying for sexual harassment legislation. Women from APC helped Sakshi to put together briefing documents for politicians and the public by undertaking Internet research on their behalf. The APC also publicized Sakshi's cause on numerous mailing lists, with the caution that supporters must be considerate of the group's restricted ICT access: "Please be aware that the SAKSHI organization is hanging on the end of a very high-cost email linkup that is *not* Internet. While they welcome your constructive input and queries, please do not begin sending lengthy documents to them without first making sure with them that they are willing to receive what you send" (APC). These efforts developed into a coordinated international campaign to promote Sakshi's objectives. In 1997, the Supreme Court of India passed a landmark ruling on a writ filed by

Sakshi with help from the APC network. Sakshi reported that the decision "...laid down guidelines to obviate such harassment at places of work and at other institutions...Most significant, [*sic*] the Supreme Court has brought sexual harassment within the purview of human rights violations" (ibid.).

Another example is the work of Bal Rashmi, a NGO based in Rajasthan, one of the poorest areas of the Indian subcontinent. Bal Rashmi's advocacy work has publicized women's experiences of rape, torture, and dowry death. When the Indian government arrested the group's leaders and filed bogus criminal cases against them, other members sought help through the Internet. Women's networks and human rights organizations were contacted in India, South Asia, Europe, and the United States. Many responded to the appeal, and actions of protest were planned and coordinated online. Within a few weeks, faxes, and letters demanding the release of the jailed activists poured into the offices of central and state governments and the National Human Rights Commission. The pressure was maintained until an investigation was announced and the cases eventually quashed (Amnesty International, 2000).

Likewise, the support of virtual networks was instrumental to the survival of Shirkat Gah, a women's group based in Pakistan that has been threatened with censure. They were one of the very few women's groups in the country that enjoyed Internet access, and they were able to forge links of solidarity through such Web sites as the GREAT network in East Anglia, United Kingdom. Shirkat Gah had caused official discomfiture by their campaign for the prosecution of honor killings (Ahmed, 2002). Allegations were made in the press by the Punjab police and the state authorities that they had embezzled millions of dollars from the World Bank, and official moves were made to close the group down. The virtual networks with which Shirkat Gah had been corresponding were quick to rise to their defense. The project evaded closure and has since substantially expanded its Internet presence and international networking partners (Moghadan, 2005).

ICT provides women with an outlet for freedom of expression in countries where dissidents are not just persecuted, but routinely executed. For instance, the Revolutionary Association of the Women of Afghanistan (RAWA) was officially proscribed under the Taliban government, but the group remained clandestinely active. The members set up a Web site in 1997 to share stories of the suffering of Afghan women with a global audience (Brodsky, 2003). The women have been audacious in their attempts to raise consciousness about human rights abuses. For example, a request to international supporters was posted on the site for small cameras that could be easily concealed under veils, in order to capture images of public executions and whippings by the Taliban. Hundreds were sent in from all over the world. The footage was uploaded on the Internet and thence broadcast on television news bulletins around the world (Benard, 2002). For the courageous women of RAWA, the Internet was a crucial resource that helped to counteract their otherwise total isolation. Indeed, as the Taliban did not allow journalists in Afghanistan, the RAWA site was often the only unauthorized source of information available on local conditions. The Web site is still operative today, but conditions for the activists remain dangerous in the context of fundamentalist resurgence.[1]

It is worthwhile to conclude this overview with the case study of ModemMujer, which integrates the themes of the Beijing conference and Mexican-based civil society (themes that are explored in the subsequent sections respectively). ModemMujer (Modern Woman) was established in 1995 as a joint initiative of five women's organizations in Mexico City. It aimed to improve communication and facilitate information exchange between urban and rural areas. During the Beijing conference, a three-woman team from ModemMujer sent regular e-mail updates to women's NGOs and individual activists; in turn, these supporters circulated the information to a wider audience via e-mail, fax, radio, print, and other media. The responses of numerous NGOs and women to developments in Beijing were filtered back to the conference team by reverse process. As a direct result, ModemMujer could refine its positions, redirect its lobbying efforts, and communicate its findings to the official Mexico NGO delegation. ModemMujer illustrated how ICT can enhance a wider, more traditional communications strategy, even in the context of poor national infrastructure. Their success was such that in the NGO delegate's post-conference report, ModemMujer were endorsed as the official communicators for the follow-up conference.[2]

ModemMujer have since helped to redefine women's political activism in Mexico by exploiting these international contacts to fight local causes. For example, in 1999 a groundbreaking public tribunal was held in Mexico where individual women collectively filed suit against the state's health care apparatus, with the support of ModemMujer. The charges included involuntary sterilization and grievous medical malpractice resulting in infant or maternal death. The event was publicized through local LaNeta conferences (the Mexican Internet provider), and the search for potential claimants was conducted by a tribunal Web site. ModemMujer's analyses of the developing proceedings were regularly distributed though their mailing list, which garnered a wide Latin American audience (E. Smith, 2000). The expanded international network was sustained after the conclusion of the proceedings through a regional discussion forum about women's reproductive health (APC).

ModemMujer has maintained a high profile Internet presence, hosting a large archive of material on women's issues and facilitating communication channels for international women's conferences. They have been nominated for prestigious international awards, in recognition for their achievements in raising awareness of the importance of ICT from a gender perspective (e.g., Global Knowledge Partnership Award 2003, Development Gateway Award 2005). They are now exploring outreach strategies to help other individuals and organizations in different regions to adopt similar models of operation. One of the women behind the project has expressed her hopes for the future thus:

> [i]t would be wonderful if other women throughout the world could initiate a project like Modemmujer, and perhaps through our voyage through cyberspace we will find each other and join together to transform a cybernetic feminist dream into a worldwide network of civil society women's organizations fighting for women's rights. (Alegre, 2003)

The activities of ModemMujer and other South-based women's groups demonstrate that online participation in the international women's movement may be wider in scope than seemingly suggested by the figures on disproportionate global Internet access. As will be seen, the rise of ModemMujer and other women's groups have been assisted by the promotion of virtual networking through the UN Beijing Conference.

6.4.2   The UN Beijing Conference

The United Nations Fourth World Conference on Women, held in Beijing in September 1995, was a watershed in the international women's movement for two main reasons. First, the events surrounding the conference provided activists with experiences of personal growth, which they translated into politicized actions in their intimate relationships and throughout their communities. Hsiung and Wong describes the impact of Beijing for many Chinese participants: "...the chance to enter the international arena provided individual women with an opportunity to forge new perceptions about themselves, renegotiate their marital relationships, and/or challenge those societal norms and practices that once governed their everyday lives" (Hsiung and Wong, 1998: 476). Moreover, women maintained the connections they had made with others during the course of the conference, often forming long-lasting international networks of political activism. Indeed, UN conferences from the Decade for Women (1976–1985) onward have had a catalytic effect on the transnational women's movement (Stewart, 2001: 227). In a study of women's participation in major UN conferences since the 1970s, Keck and Sikkink observe that these encounters "generate the trust, information sharing, and discovery of common concerns that gives impetus to network formation" (Keck and Sikkink, 1998: 169).

Second, more than any other previous conference, Beijing heralded the arrival of the Internet era (Dickenson, 1997: 109). Harcourt quotes leading participant, Alice Gittler, thus:

> [e]lectronic communication allowed women to bypass mainstream media and still reach thousands...women who met online found an immediate network...One hundred thousand visits were made to the APC website on the Conference...When the International Women's Health Tribune Global FaxNet was posted on the web, over 80,000 hits were recorded in the week before the Beijing meetings. (Harcourt, 2000: 54)

Indeed, significant strides made been made toward building a transnational network before the conference even began. One of the most active discussion forums was Beijing95-L, which included topics such as pre-conference events, NGO information, formal and informal reports on women's issues, and job and volunteer opportunities related to the conference. The discussion forums hosted by the UNDP, such as Womenwatch, also attracted thousands of postings from hundreds of delegates and other interested parties across the globe. For example, the UNDP sponsored Beijing-Conf, a popular listserv attracting subscribers from 55 countries, including 28 underdeveloped states—an unusually high proportion. In addition, there was an official UN Women's Conference Web site and alternative sites such as NGO Forum Daily, the U.S. NOW, WomensWeb Canada, and Virtual Sisterhood.

The online conversations were not just conference-specific, but also encompassed wider debates about political activism and the future direction of feminism. Dialogue continued long after the conference ended. Participants expressed genuine excitement about the potential of the Internet to aid transnational mobilization, as the following extract from a post on the UNDP Women-Rights list illustrates:

> We can organize across national boundaries and really have an impact. The Internet offers a way to communicate with one another and coordinate our efforts. Surely if tens of thousands can go to Seattle to demonstrate against WTO, then hundreds of thousands of us can use the Internet—and our buying power—to make our views felt in corporate headquarters of companies exploiting women. (Women-Rights List Archive, c)

The Internet elicited enthusiasm among activists because it represented an interactive and flexible alternative to mass media—aspects of which have been long-standing targets of feminist critique (e.g., Douglas, 1995; Valdivia, 1995; van Zoonen, 1991). As a NGO expert observed in a post on the UNDP Women-Media list:

> Women are using whatever media are most useful and appropriate to communicate and exchange information at any given time or place: songs, e-mail, posters, the web, poems, video, plays, magazines, radio, drawings...However, the so-called "traditional" media, which in many places are really the "mainstream" media...are often used to misinform, distort and disempower women...Over the years I have also seen many women's groups successfully use new technologies for information sharing and communication. (Women-Media List Archive, a)

These sentiments were echoed by another activist, who urged closer cooperation among fellow participants for a common cause:

> We women must create alternatives in different media and use them to inform and empower women, to get women out of their isolation. We must make ourselves more visible and audible so that our concerns do not remain unarticulated and unattended. Not only must we evolve alternative messages but alternative methods of working together; methods which are more democratic and participatory and which break the divide between "media makers and media takers." (Women-Media List Archive, b)

This exemplifies many similar postings that explored a media-centric approach to women's empowerment, in which new media, and especially the Internet, were identified as vital resources. The above appeal for grassroots involvement was also echoed elsewhere in the conversation thread. The topic received extensive and detailed dialogue about appropriate social and political strategies to attain these goals (Women-Media Archive).

Online dialogue instigated by Beijing has contributed to highly productive networking that has extended well beyond the duration of the conference (Steans, 2003: 89). Through the UNDP Web site, disparate women's groups have been able to exchange advice about how to best adapt to the new media environment, which has wider benefits to the women's movement as a whole. What is evident from all the extracts above is a common determination to take advantage of the opportunities afforded by ICT to bolster the transnational network of women activists. This commitment seems to be indicative of feelings of mutual affinity, despite the rich social and cultural diversity of the participants. Luo Xiaolu, another participant, told of how the conference increased her awareness of the importance of the type of cross-cultural communication that virtual networks can facilitate:

> ...I realized that many aspects of women's issues transcend national boundaries and other differences. Women throughout the world face many common problems. However, we have little opportunity to communicate, especially between China and foreign countries. In China we talk about "women-work" [*funu gongzuo*], while women elsewhere speak of the "women's movement" or "feminism." This difference is caused by lack of communication. So, I feel the first thing we need to do is communicate, among ourselves and with women abroad. Even though people many have different ways of doing and thinking things, or may even have different starting points, our goal and objectives are the same. (Hsiung and Wong, 1998: 483)

Since Beijing, women's transnational networking has continued to proliferate, particularly in regard to the recent Beijing+5 Review Conference. The International Women's Tribune Center (IWTC) have capitalized on the gains made at Beijing by setting up a Web site for the five-year review, with contributions from a global coalition of feminist Web sites, and in cooperation with WomenWatch, the UN women's Internet resource.[3] Discussion lists are still active regarding the implementation of the Women's Conference and Social Summit agreements, and other post-regional and post-conference follow-ups. In addition, the WomenAction Web site was established as a "global information, communication and media network that enables NGOs to actively engage in the Beijing+5 review process with the long-term goal of women's empowerment," available in English, French, and Spanish.[4] Access to Internet resources now seems to be widely regarded as a *necessity* for conference participants: in WomenWatch's words, "as a tool for women's empowerment" (WomenWatch).

In sum, like the international women's movement of the past, women have found imaginative and effective ways of networking across state borders to share concerns, express solidarity, and attempt to achieve common political goals. There is evidence of genuine camaraderie and examples of political efficacy. ICT have been pivotal in the recent explosion of collaborative activity, which includes linkages between high- and low-income countries, large NGOs, and individual activists, and the grassroots and global governance.

6.5   The Zapatista Movement

The indigenous Zapatista rebellion in the southern Mexican state of Chiapas has developed into a fascinating case study of the capacity of ICT to assist in mobilizing transnational support for localized politics (Holloway and Peláez, 1998). It is an example of what Routledge terms "globalised local action," where "the virtual geography of the [I]nternet becomes entangled with the materiality of place, local knowledge, and concrete action" (Routledge, 2001: 31). The Zapatistas have built upon local initiatives to create transnational networks of communication, solidarity, information sharing, and moral support (Routledge, 1998). The success of this strategy has been extraordinary.

Ironically, electricity and telephones are extremely scarce if available at all in the rural areas of Chiapas, much less Internet access. Yet this unlikely setting was the origin of what has been popularly termed the world's first "postmodern revolution" (Burbach, 1994). Pro-Zapatista solidarities have reached across five continents and dozens of countries, generating the rapid growth of much wider activism (Olesen, 2004). Moreover, support for the rebellion has evolved into a "kind of electronic fabric of opposition" against neoliberalism (Cleaver, 1998b). An authority on the movement, Cleaver boldly asserts that the Zapatista networks "are providing the nerve system of increasingly global challenges to the dominant economic policies of this period and in the process undermining the distinction between domestic and foreign policy and even the present constitution of the nation state" (Cleaver, 1998a: 622). Perhaps an ambitious claim—but the Zapatista blueprint for resistance was certainly inscribed with fundamental challenges to traditional conceptions of political boundaries. The formula has since been endlessly imitated by other social movements.

The Zapatista Army of National Liberation (Ejército Zapatista de Liberación Nacional or EZLN) staged an uprising on January 1, 1994; the day that the North American Free Trade Agreement (NAFTA) came into effect in Mexico. The action was a show of protest against neoliberalism and a demand for greater self-determination and more equitable distribution of resources. A military conflict ensued, and the Zapatistas managed to claim some territory before being routed by the Mexican Army a year later. The Zapatistas retreated into the jungles of Chiapas, where they were the targets of years of low-intensity warfare. They eventually foreswore militarism and sought international support through a series of public declarations. There has subsequently been sporadic talks with the government, most notably culminating in the San Andrés Accords in 1996. The Zapatistas claim that the government did not honor their commitments, and that they and the indigenous population has suffered an officially sponsored campaign of repression ever since.

The electronic dimension to the Chiapas rebellion began as a counterpropaganda response to the Mexican government, which initially presented the suppression of the uprising as necessary measure to preserve national integrity. The Zapatistas were able to effectively rebut this claim by communicating with the outside world via e-mail bulletins. The Zapatistas protested that far from conspiring to overthrow the state, they only sought indigenous autonomy. Indeed, from the earliest stages, the rapid dissemination of information through ICT was a vital tactical advantage of the pro-Zapatistas (Russell, 2001b). The Internet enabled the activists to contest erroneous reports in the media and to issue speedy rebukes to government disinformation, often influencing mass media coverage in favor of the Zapatistas.

The successful execution of this media strategy was impressive considering the extreme deprivation of the indigenous peoples. In general the Zapatista movement had a mediated relationship to the Internet, notwithstanding some direct input from senior lieutenants. The hyped-up image of troop leaders using "a portable laptop computer to issue orders to other EZLN units via a modem" has no basis in actual evidence (U.S. Army, cited in Swett, 1995). In the widespread absence of electricity, the rebels did not often have a direct online presence, and instead relied on a wider supporter network to disseminate their message. In addition, written communiqués were given to sympathetic journalists for publication. These materials would subsequently be scanned or typed up for online distribution by pro-Zapatistas (Russell, 2001a). Activists also helped to create a number of specialized Web pages and discussion lists, specifically dedicated to tracking the situation in Chiapas. The sites hosted firsthand observation reports and specialist analytical commentary that was often cited by news reports on television and radio. Senior lieutenant Subcomandante Marcos evaluates the success of this approach:

> There are people that have put us on the Internet, and the *zapatismo* has occupied a space of which nobody had thought. The Mexican political system has gained its international prestige in the media thanks to its informational control, its control over the production of news, control over news anchors, and also thanks to its control over journalists through corruption, threats and assassinations. This is a country where journalists are assassinated with a certain frequency. The fact that this type of news has sneaked out through a channel that is uncontrollable, efficient and fast is a very tough blow. The problem that anguishes Gurria [secretary of foreign affairs] is that he has to fight an image he cannot control from Mexico, because the information is simultaneously everywhere. (Froehling, 1997: 297)

Subcomandante Marcos neglects to acknowledge the importance of his personal magnetism in attracting keen international interest. His literary writing style, media-savvy, and mysterious image combined to make him something of a counterculture celebrity. He has been interviewed by a variety of media outlets, and portrayed as a latter-day leftist icon (De Huerta and Higgins, 1999). The centrality of Marcos in the Zapatista discourse can be critiqued as symptomatic of the "cult of personality" that is often thought to afflict establishment politics and the mainstream media. According to the public sphere perspective, excessive personalization of political issues can distort the conditions for rational-critical discourse. Marcos' renown appears to be partly a media construction, and partly a reflection of genuine admiration from supporters who were quick to adopt his image as symbolic of the movement.

However, perhaps the most critical factor behind the widespread appeal of the Zapatistas was their refusal to be defined by any traditional political ideology, which struck a chord with many activists disillusioned by the fate of socialist politics in the past century. The Zapatistas present little more than "proposals for action" to the outside world rather than detailed manifestoes. Their main goal is political autonomy, but they purposely omit the exact framework and details of their ideas in order to foster dialogue (New Internationalist, 2001). The global audience is explicitly invited to join the conversation. Thus, the inaugural list of Zapatista demands was as expansive and nonspecific as possible, calling for "work, land, housing, food, health care, education, independence, freedom, democracy, justice and peace" (EZLN, 1994a). In addition, the "Woman's Revolutionary Law" was a particularly provocative declaration in the context of Mexican patriarchy (Autonomedia, 1994). This list of demands for women's rights was portrayed as emblematic of the Zapatista commitment to inclusiveness. It reflected the prominence of women in the Zapatista army—a full one-third of the total.

The obstacles to the Zapatista's goals were described as an economic and political clique that "...don't care that we have nothing, absolutely nothing, not even a roof over our heads, no land, no work, no health care, no food nor education" (EZLN, 1994a). The rebels dismissed the government's responses to their demands as "...a series of offers that did not touch the essential point of the problem: the lack of justice, liberty and democracy in Mexican territory" (EZLN, 1994b). The political establishment was characterized as a "...system of complicity which makes possible the existence and belligerence of *cacicazgos*, the omnipotent power of the cattle ranchers and businessman and penetration of drug traffic..." (ibid.). What is interesting is that within a short space of time Subcomandante Marcos publicly redefined the enemy as international neoliberalism, because

[w]hen we rose up against a national government, we found that it did not exist. In reality we were up against great financial capital, against speculation and investment, which makes all the decisions in Mexico, as well as in Europe, Asia, Africa, Oceania, the Americas—everywhere. (Starr, 2000: 104)

The Zapatistas adopted the rallying cry of "*Ya Basta!*" (Enough!) as a slogan, that symbolized their rejection of neoliberalism and economic globalization. They claimed the oppressed masses worldwide as natural allies in the Zapatista struggle. In the past, Zapatistas have expressed support for homosexuals, ethnic minorities, indigenous peoples, and environmental activists. Subcomandante Marcos expressed the spirit of Zapatismo (and made a pitch for international support) by declaring "...in San Francisco, Marcos is gay...Marcos is all of the minorities who are untolerated, oppressed, resisting, exploding, saying 'Enough'." (105) These appeals to solidarity and common identity are typical of EZLN material (Stahler-Sholk, 2001). The declarations were widely circulated online, and became staples of the Web sites managed by the self-styled "intercontinental cyberspace liberation fleet' behind the EZLN. In an interview with e-zine "Wired

News," Tamara Ford, joint administrator of Accion Zapatista and ZapNet Collective, assessed the effect of Zapatista publicity:

>...there is a larger Zapatista discourse—reflecting a very profound commitment from the indigenous communities willing to put their lives on the line—that most people don't get to see. It's not printed in our newspapers. That's why the Net's been so important in distributing information that allows people to go beyond any romantic limitations of the left. Moreover, most of the Zapatista supporters are engaged in their local struggles, which they see as very connected to what the EZLN is fighting for. Thus, the idea of the "other" is collapsing. We are one. (Wired, 1998)

These sentiments can be interpreted as indicative of a sense of mutual affinity amongst the virtual Zapatista network, based on shared experiences of oppression under conditions of liberal economic globalization. The network was broad-based, evolving from others already extant, such as Latin American and indigenous networks and anti-NAFTA groups (Johnston and Laxter, 2003; Olesen, 2004). But it also eventually evolved to include representation from all of the minorities to which Marcos pitched his appeals.

Political mobilization was remarkably swift. Within days of the paramilitary repression, human rights organizations appealed for donations and volunteers for humanitarian observation, and besieged the Mexican and U.S. government with letters and e-mails of protest. Mexican voluntary groups worked with their American counterparts to bring human rights caravans from the United States to Mexico (Cleaver, 1998a). Activists also exploited the Internet to rapidly coordinate "virtual demonstrations." For example, a report regarding a leaked memorandum from Chase Manhattan Bank urging the Mexican government to deal decisively with the Zapatistas was circulated widely online and led to protests about the government acting at the seeming behest of international capital. In a posting to an online discussion list, Harry Cleaver describes how quickly and effectively the community of international activists was able to mobilize in response to this call for action:

>The Chase Manhattan report to emerging investors, written by Riordan Roett was on the Net when Ken Silverstein called me up and told me about it. He faxed me a copy which I typed up into e-text and posted. The extremely rapid circulation of that report resulted in widespread mobilization in the US against Chase. It was one of those rare moments of frankness that just happened to fall into the hands of those

for whom "investment" in Mexico means support for democracy and indigenous rights, not profit-making. We made good use of it to illustrate the forces behind the military's actions. (Wired, 1998)

Internet activity was supplemented by usage of other ICT and traditional campaigning methods. For instance, Internet protests forced Televisa, Mexico's largely state-controlled TV network, to report the official demands of the Zapatistas at crucial moments in negotiations with the Mexican government (Halleck, 1994). Another example is that organized demonstrations held in the United States outside Mexican consulates in 1997 were supported by Zapatista-sympathizers worldwide, who bombarded the consuls with faxes. Tamara Ford describes how protest unfolded in the interface between different ICTs, following a wave of paramilitary killings of Zapatista supporters and civilians in Chiapas:

> There have been letter-writing campaigns and forms of virtual protest. Chiapas95 has distributed hundreds of reports from demonstrations in dozens of countries in recent weeks. There were various proposals for coordinated Net action, including a Net-Strike targeted at the servers of Mexican Financial Centers. Another proposal that circulated on the Net was a project to provide indigenous communities with video equipment and training to document human rights abuses. This project actually got under way in Chiapas within weeks but its director was promptly and illegally deported by the Mexican government. News of *this* development also circulated with great speed. (Wired, 1998)

The superiority of Zapatista Web resources compared to the Mexican government was notable throughout the early years of the uprising. The state had two different official pages in operation and many ministries failed to update their information for more than six months at a time (Cevallos, 1998). Poorly maintained government sites were an embarrassing contrast with EZLN sources, which boasted swift navigation, attractive design, and constantly updated information. But the Zapatista network would prove a significant challenge for most governments in this regard. They had the honor in the late 1990s of being commended by Wired magazine for having the best-organized and most dynamic Web presence of any political group anywhere (Wired, 1998).

The Mexican government has publicly acknowledged the role of the Internet in undermining its credibility and challenging its policies. As early as April 1995, Mexican secretary of state, Jose Angel Guru, was publicly defining the conflict in Chiapas as a "war of ink and Internet" (Froehling, 1997: 304). The government's counterinsurgency efforts have been muddled and largely ineffective. There is widespread suspicion that the government has attempted to sabotage the rebel's computer link at certain intervals. For instance, LaNeta (the Mexican Internet provider) went down immediately prior to the posting of the Zapatista's declaration of demands, and "worried communicators flooded several conferences with accusations of government censorship" (Halleck, 1994). The information was swiftly rerouted through another service provider. Regardless of whether there was any substance to the conspiracy rumors, the incident illustrates the difficulties surrounding official censoring of the Internet.

Despite government attempts at repression, Zapatista sites continued to proliferate. Hundreds of Web pages still exist today with detailed information on the Chiapas situation. Numerous sites and discussion lists circulate daily updates of information.[5] Some projects originated in Mexico such as the list Chiapas-1, which was run through the UNAM (Universidad Nacional Autónoma de México, or National Autonomous University of Mexico) computers in Mexico City, or the FZLN-1 list, which was run by the Frente Zapatista de Liberacion Nacional (FZLN, or the Zapatista National Liberation Front). The origins of others lie outside of Mexico: for example, the first unofficial EZLN Web page was run through the Swathmore Web server in Pennsylvania, United States. There are also major Zapatista sites hosted in Austin, Texas,[6] and Dublin, Ireland.[7] "Radio Insurgente" also provides dedicated online streaming of EZLN broadcasts.[8]

The activist network has fragmented across many platforms. The diffusion illustrates the inevitable social segmentation that results from a diverse transnational realm, which is further encouraged by the "many-to-many" networked structure of the Internet. Even so, what began as an interlinked set of spontaneous actions has evolved into something more organized over time. For example, certain sites such as MexNews operated a cooperative division of labor where a group of subscribers took individual responsibility for uploading and posting relevant material to other sites and lists such as Chiapas-1 and Mexico2000. Another site, Chiapas 95, was specifically set up to reproduce the very best material from these postings to provide an effective summary for casual observers or those pressed for time.[9] Skills, resources, and information were pooled in order to benefit the group of subscribers as a whole. The presence of mutual affinity seems evident.

The movement has also invested in independent media production, to promulgate their message and preserve their stories for posterity.

There are several examples. Collaborative efforts were undertaken to produce an electronic English translation of the only existing archive of interviews and documentation about the women of Chiapas. A collective of activists produced a multimedia Zapatista CD, which provides an overview of key events through archived Internet material and audiovisual clips (Cleaver, 1998b: 7). An electronic book *Zapatistas! Documents of the New Mexican Revolution* was put together by a team of supporters via e-mail, using translations of material largely gathered from Chiapas discussion lists (Autonomedia, 1994). It has been made available online as well as in hard copy. To encourage the widest possible dissemination of the book, the authors have decided not to claim copyright protection. In addition, pirate video have been circulated for "teach-ins," pro-Zapatista interviews, and indigenous music have been compiled on audiotape and CD, and pirate radio stations and community-access TV has been used to counter biased coverage in the state-controlled media (ibid.). Alternative media sources have been important in raising awareness of the Zapatista cause among those who are unable to access the Internet, such as the poorer classes in Latin America.

The Zapatista sites have entered discussion about establishing an Intercontinental Network of Alternative Communication, for groups with an interest in Chiapas and anti-neoliberalism. Hence, on-site discussion has moved from a specific event in time—the indigenous uprising—to broader issues of capitalism and the state (Gallaher and Froehling, 2002). The Zapatistas have explicitly encouraged this development, which suggests that the electronic networks that have grown around the EZLN are not a temporary coalition. In fact, activists have sought to consolidate transnational networks by organizing opportunities for face-to-face interaction. In August 1996, the Intercontinental Encounter for Humanity and against Neoliberalism was held in Chiapas as a symbolic show of solidarity with the rebels. This event, which had been preceded by similar encounters in Europe, America, and Asia, brought together activists from all over the world in a show of protest against "neoliberalism," in the name of "humanity." There were 3,000 attendees from 5 continents and 42 countries (Cleaver, 1998a: 627). Since then, the conference has become an annual event and an important fixture in the calendar of the global anticorporate movement. It serves to reinforce the wider international implications of the Zapatista struggle as well as to maintain domestic political pressure about indigenous demands for reform (Routledge, 2001: 26). Wider media attention has been aroused by the presence of political and celebrity figures at past conferences, such as U.S. film director Oliver Stone, Danielle Mitterrand, widow of the late French president, and celebrated French activist José Bové. Newspapers that regularly cover the event, such as *The Guardian*, have marveled on the extraordinary transformation of the Zapatistas from a small guerrilla army to "an international cultural movement...emphasized by the remarkable mixture of supporters flocking the streets, from the capital's smartly turned out bourgeoisie to body-pierced and pink-haired punks...and dungareed Italian anti-globalisers" (Campbell, 2001).

The 1994 uprising created a flurry of international interest and presented the Zapatistas with a difficult challenge: how to increase and maintain a transnational coalition of supporters. They have met this challenge with such spectacular success that they retain a high profile in global civil society almost 15 years after they first rose to prominence. Astute use of ICT, even in difficult circumstances, has helped the Zapatista movement to magnify their sociopolitical impact. But just as important were explicit entreaties for global dialogue and normative appeals for solidarity. The Zapatistas recast their struggle from a localized dispute into an iconic struggle against prevailing forms of global repression, thus creating a basis from which to form bonds of affinity with marginalized others.

6.6 GREENPEACE

There is a gathering sense of crisis about the global environment. We are daily confronted with disturbing evidence of humankind's detrimental impact on the planet, from the effects of climate change to mass species extinction. Environmental issues have moved from the periphery of the political agenda and become an increasingly central preoccupation of international politics. The green lobby has been a key driver in this process, and ICT has been instrumental in the formation of cross-border environmental networks. In his account of emerging global environmental governance, Lipschutz cites numerous examples of influential networks, such as the Climate Action Network (CAN), the Global Rivers Environmental Educational Network (GREEN), and the River Watch Network, as well as individual campaigns such as over the Mattole watershed in northern California, the Amazon rainforest, Love Canal, and the residents of Owens Valley in eastern California against Los Angeles (Lipschutz, 1997). Keck and Sikkink's study of environmental transnational advocacy networks details the intricate cooperation behind the multilateral development bank campaign and the campaign against deforestation in Sarawak (Keck and Sikkink, 1998). There is a growing body of literature about the activities of the international environmental movement that acknowledges the importance of ICT in achieving campaign goals (e.g., Mancusi-Materi, 1999; Pickerill, 2003; Wapner, 2002). These examples illustrate that even when environmental degradation is localized, electronic networking can amplify resistance until it is global in scope. Governments and corporations have to increasingly anticipate these possibilities in their political and business strategies.

Before proceeding, it should be noted that some groups criticize computer technology as ecologically detrimental, lest the impression is given that there is a consensus amongst the wider movement that ICT is an intrinsic good. Glosserman outlines some of the environmental consequences of computer production:

> Production of a single PC requires 33,000 liters of water, generates 290 kg of waste and consumes 5,000 kwh of energy. Average use during one year sucks up 85 kwh, although being attentive and careful can cut that figure to about 40 kwh...Take the entire production process, including the mining of the rare metals and the manufacturing of chips, and you've got a lot of petrochemicals and silicon in that box on your desk. (Glosserman, 1996)

There has nonetheless been an inexorable rise in virtual networks of environmental activists, with evidence that most perceive the Internet as having a less harmful impact on the environment than paper production (White, 2000).

One of the most well-known and successful environmental NGOs is Greenpeace. In recent years, it has made pioneering use of the Internet as a tool in environmental campaigns. For many environmental organizations, Greenpeace has become something of an exemplar in effective "cyberactivism." Greenpeace's first basic Web page was set up on August 30, 1994. Progress was rapid. Within less than a month a major ozone campaign site had been established. Within a year, it had publicized a secret nuclear shipment route from France to Japan and appealed to activists to fax the French Embassy and write letters of protest to the newspaper, Le Monde (Greenpeace, b). The French government was reportedly so overwhelmed with faxes that it demanded that Greenpeace remove the fax number from its site.

Five years later, Greenpeace had grown more sophisticated in their methods. In June 2000, activists installed a Webcam at the end of an underwater radioactive discharge pipe operated by the French nuclear agency Cogema, in la Hague, France. The Webcam provided live feed of nuclear waste discharge that was simulcast on the Greenpeace site and a screen at the Convention for the Protection of the Marine Environment of the North-East Atlantic (OSPAR), where delegates were discussing the future of nuclear reprocessing (Greenpeace, a). Greenpeace included a facility for visitors to the site to post messages to the OSPAR delegates, which were then broadcast on the screen. The site was swamped with contributions from nearly 2,500 people. The situation took on a surreal dimension when Cogema sent down divers to display a banner before the Webcam, which claimed that the discharges had "Zero Impact." Nevertheless, the North-East Atlantic countries called on France and the United Kingdom to end their nuclear reprocessing (Brown, 2000; Parmentier, 1999).

Currently, Greenpeace has perhaps one of the most extensive and well-maintained NGO Web sites in the world.[10] It was recently supplemented by a dedicated multimedia resource called the "Cyberactivist Center," as an extension of its conventional campaign work.[11] The coordinator estimated that within three years of inception, the Center reached around 100,000 registered cyberactivists from 216 countries (in addition to about 200,000 cyberactivists registered by their national offices). The activists participated in 400,000 action alerts that resulted in at least 800,000 e-mails and faxes to corporations and governments. Around 180,000 e-cards promoting Greenpeace were sent by activists to friends and colleagues, and the discussion list received 22,000 postings, ranging from short comments to lengthy articles. At its peak, membership was growing by more than 100 people a day (Greenpeace, a). The Cyberactivist Community claims a number of campaign successes. Examples include pressure for a referendum against loading plutonium into a Japanese nuclear reactor; publicizing destructive logging in Canada's Great Bear Rainforest; securing the end of a major source of genetically engineered corn in Mexico and significant restrictions on GM foods in Brazil, Europe, China, and elsewhere; bringing logging on Dene land in the Amazon to an end; and winning a permanent extension on a moratorium on mahogany logging. Developer of the center, Keith Jardine, describes its contribution to Greenpeace's activism:

> The centre provides a cyberactivist community where people representing over 170 countries and territories can share ideas and participate in environmental actions such as the recent Corporate 100 actions against global warming [a campaign to pressurize the 100 largest US companies to support the Kyoto Protocol]. With a simple mouse click, cyberactivists can learn which US companies oppose the Kyoto Protocol and can take that information into the real world to use their buying powers accordingly. (ibid.)

The Corporate 100 campaign to which the quote refers employed a number of colorful methods to engage the broader public in the issues, such as interactive games and humorous flash-animated e-mail postcards. An action kit was also available to download, including posters, leaflets, and suggested ideas for newspaper articles—all designed to encourage activists to run their own local public campaigns. The kit was accessed by over 4,000 activists (ibid.). These campaigns are heavily dominated by members from the most developed countries—where most of Greenpeace's supporters and donors reside, and where most of the world's Internet users are located. But Greenpeace have also made attempts to extend "cyberactivism" to the global South. In 1999, Greenpeace employed its first "worldwide media campaigner," Hemant Babu, in India. Babu's first action was to establish an Internet cafe in front of the abandoned Union Carbide factory, scene of the notorious 1984 Bhopal disaster. Thousands of Bhopal residents visited the cafe to send e-mails to the Indian government, Union Carbide and Dow Chemicals, demanding for action on the leaching of toxic chemicals into local groundwater. The issues with Union Carbide persist, but the intense pressure that the activists were able to exert was evidenced by the company's decision to screen out Greenpeace-sourced e-mail (ibid.; also see cyberactivist discussion at Cybercenter Archives, b). Babu considers efforts to expand access to ICT in LDCs to be an essential element of Greenpeace's work:

> This form of cyberactivism empowers people with modern information technology to fight their battles against environmental criminals. Given their socio-economic conditions it is too much to expect that they will come to technology. So our job is to take technology to them…By doing this we ensure that information technology does not remain an elitist concept but reaches the most oppressed strata of society, which is also often the worst affected by environmental degradation. (Greenpeace, a)

On its well-subscribed discussion forum, the Cyberactivist Center often sought feedback on its progress. One year into the establishment of the Web site, Radagast, the main coordinator, reflected on how the development of the site had compared with initial expectations:

> When we launched the site a year ago, it was very much an experiment… I really had no idea what would happen. I had two fears, however. One was that the site would be ignored—we would get no participation and no discussion. That certainly has not happened. We've been averaging

some 40–50 postings a day, and more importantly, many hundreds of people a day participate in some action—faxing a letter, sending an e-card, downloading an action kit, amongst many other options...My second fear was that the site would be quickly dominated by flamers, lamers and trolls, the bane of any discussion centre. I worried that people would be posting insulting comments, foolish comments and irrelevant material and the site would fail to achieve its main purpose—to help us win campaigns. This has also not happened. Although there are occasionally heated discussions and rude comments, by and large the postings have been interesting, insightful, and occasionally even profound. A lot of good campaign ideas have been generated, friendships made and even some lives changed by the discussion that has taken place here in the last year. (Cybercenter Archives, c)

These claims sound like hyperbole, and one expects the coordinator to put the most positive spin on Greenpeace's projects. However, the post was substantiated by numerous messages of strong support from cyberactivists worldwide. The extracts below give a sample of direct replies on the listserv to Radagast's post. A remarkable conversation developed between the members that reflected their perceptions of the personal incentives and political benefits of cyberactivism:

Congratulations Radagast and all others at the Cybercenter for an excellent and productive year. It was about this time last year that I found this community. It was indeed, a life-changing experience...

A year ago, I honestly didn't know what I was getting into...But now, a day doesn't go by where I don't check the boards and roam other sites. I have been writing much less than before because I've been so busy planning and actually executing those plans! Just like this site, I too have evolved.

I...would just like to thank Greenpeace for the initiative of "Cyberactivism." A big problem with environmental work is that many people feel that they don't really have the time to act, but with this great webforum many, including myself, can help to protect and save the environment. Thank you all!

All these victories [campaign victories relating to the Cybercenter] were on the back of years of campaign work in the public and political worlds, but the core message we keep sending is clear: the cyberactivist world is watching.

I do see a place for sites like this in getting activists involved as a first step...but I see it as a spring board to learn from...and once your levels of comprehension of the issues are clear individuals are able to then launch themselves into real activism, on the streets, not only in stuff like marches and rallies, but in "local issues" group forming, growing and facilitating...(ibid.)

The quotes illustrate the high value that many place on Cybercenter membership; indeed, the authors of the first and second extract explicitly attribute involvement with cyberactivism with a sense of personal growth. The third extract identifies the most appealing aspect of cyberactivism as convenience in the context of a busy lifestyle, which is typical of opinion expressed in the ensuing dialogue (ibid.). However, the fourth and fifth extract makes the point that the efficacy of cyberactivism is dependent on complementary forms of traditional campaigning. The virtual network is not divorced from the "real world" network of Greenpeace activism; rather the two are inextricably linked. The translation from online to offline activity is a topic that particularly interests Greenpeace organizers and activists, and it has been repeatedly debated. Some posts from members have signaled dissatisfaction with the limits that virtual networking imposes. For example:

> For me cyberactivism is the only way that I can be involved. I think that it is a great idea and I have tried (since I discovered this) to involved [sic] much as I could. Of course I want to be involved physically in Greenpeace campaigns and also I want to be informed when the Greenpeace activists act in my country. I was verry [sic] sad when I saw on tv Greenpeace action in my town and I didn't know about it. I like to know these things from you, not from tv...

> To a small extent, I do feel as though I am making a difference by sending e-letters to different corporations; however, I would like to be more actively involved. If you could suggest ways on how we could do something other than emailing letters, I would get more involved in your campaigns. (ibid.)

Both posts are indicative of a wider appetite for activism that, for some, the Cybercenter is unable to satisfy. It is interesting that the author of the first extract states a preference for direct communication from Greenpeace via the Internet, as opposed to the message being channeled through broadcast mass media. The implicit suggestion of both authors is that the Cybercenter needs to improve its targeted mailings to activists, most pointedly in the second author's plea for more opportunities for participation. Radagast has responded to such posts with a keen critical assessment of the Web site's past achievements and future potential. He conceded the need for improvements and appealed to members for constructive suggestions:

> One of the main reasons Greenpeace originally set up the Cybercenter was to get some desperately needed ideas and volunteer activists for our ongoing campaigns. This has worked beautifully in terms of letter writing

and ecards...but has not worked well in generating new ideas and real-world actions that we've actually been able to put into practice to help us with our campaigns. I know that many of you want to get more involved in Greenpeace campaigns, both online and in the real world, so something is going wrong somewhere. (Cybercenter Archives, a)

Clearly this consultation was a response to demand from many grassroots members for increased participation. But as revealed in the discussion that followed, other members welcomed the opportunities that cyberactivism provides to customize the extent of their involvement:

> I have a really busy job, but thanks to the cyberactivist newsletter I manage to send about 4–5 communications a week to people engaged in environmentally and socially questionable practices. I never wrote letters. I also change my purchasing habits based on your campaign information and put up posters in my workplace, and communicate your info to all my interested friends and colleagues, so I don't think that you should see the lack of community conversation as a failing.

> As a personal anecdote, I have gone through phases of being an organizer and an activist in my community versus times when personal matters had higher precedence and immediacy. During the latter times it is all I can do to write letters and keep current with what is happening on this thankfully fast-paced site. So those of us who move between roles this site provides a mechanism for us to help mentally active, informed and involved, while allowing us to hone our letterwriting [sic], composition and debating skills.

> I live in a town in Scotland, and I know NO ONE who knows anything about environmentalism, they do not realize what a desperate state the world is in, they think it's just a lost cause "hippys" [sic] think about...(ibid.)

These three contributions exemplify the varying depth of attachment to the virtual community felt by cyberactivists. The first extract indicates a fairly minimalist approach to the Cybercenter, but owing to tight time constraints, the author evidently appreciates even this modest contact. Her contribution does not extend to online dialogue, but she participates indirectly by relaying information to her immediate social circle. A similar perspective is expressed in the second extract. Here, the author directly accredits his enhanced skills of self-expression to cyberactivism, again reflecting the theme of personal growth. The third extract typifies many contributions to this conversation thread, which described cyberactivism as a means to remedy a sense of geographical isolation. As Ward argues, members of virtual communities can derive great value from online interaction, even when their involvement is limited and episodic (Ward, 1999). It would seem an instrumental approach to virtual community does not necessarily negate meaningful experiences.

The divergences between the above perceptions of cyberactivism were reflected in interesting exchanges whereby the subscribers analyzed the usual patterns of discourse in the Cybercenter. Radagast advanced that there are three different kinds of cyberactivists: the "ziners," the "organizers," and the "discussers," (Cybercenter Archives,c). He argued that the "ziners" are the most common type: they read the monthly e-zine, the occasional update, and send letters and e-cards; however, they do not tend to read the discussion forum and never submit posts. Hence, despite their input, they are the least active members. The "organizers" tend to be very active within their local community and mainly use the Cybercenter to raise international awareness of campaigns in which they are already involved. They post articles to the forum nearly every day, and often post comments and action alerts under their own articles, rather than inviting the involvement of other cyberactivists. Organizers therefore largely use the Cybercenter insofar as it will publicize their own causes, rather than embrace the participatory nature of the discussion forum. It is the "discussers" who use the site most frequently, and who tend to engage in detailed and lengthy debates about environmental politics, Greenpeace, and the Cybercenter. As a result of this analysis, Radagast and the other Cybercenter administrators proposed a redesign of the site to cater more efficiently for each of these user categories. The sign-up process was streamlined for the ziners, and options for a more "user-friendly" HTML version of the e-zine were explored. The community forum that is popular with the organizers was improved and greater effort made to support local action groups. In addition, discussion features were extended to all parts of the Greenpeace site so that users could post comments under any published article.

Judging by the responses prompted by Radagast's post, the typology of different users was widely accepted as accurate. However, as the following extracts demonstrate, there was also some disagreement:

> Being a discusser, I guess, I'd like to know if there's really that much distinction between the actors, "ziners," "organizers," "discussers?" I think often it may be the same people in different roles or state of minds. Certainly I know that when I am typing something here "as a discusser," which I was convinced to do without a name, by the excellent arguments re. anoynimity [sic], I feel different than when my name is on something I organized. Which I do very often ... Accordingly,

I might, as a "ziner," just send a mail required without thinking too much about it. (ibid.)

The above quote illustrates that whilst distinctions between categories of cyberactivists provide useful generalizations, there are significant ambiguities regarding "role-adoption." A person can assume a number of different roles over a period of time depending on external commitments, and/or the issue-area concerned. Again, here is testimony to what is popularly perceived as a distinct benefit of cyberactivism: the ability to tailor the scope of one's involvement to suit other aspects of one's life. Participation can range from the mere clicking of a button to forward an e-mail, to the more time-consuming option of becoming fully involved in Web debates on the nature of campaigning. But also, here is evidence to reinforce Bohman's accusation that the anonymity of Internet communication can distort critical publicity. The author alludes to altering her position depending on whether she chooses to disclose her identity. Certainly, a large number of members post anonymously or under a pseudonym. Skeptics of virtual networks argue that anonymity has the effect of making the sincerity of discourse difficult to gauge (e.g., Poster, 1995a). Whether anonymity aids or hinders free speech is a moot point. If participants can effectively escape accountability for their opinions, it would seem likely to create a climate of mistrust. Yet it is also possible to identify the formation of norms specific to Internet discourse in virtual communities, which might mitigate these problems. For instance, in many forums, the use of pseudonyms has become a widely accepted form of concealing one's identity for security reasons; a pragmatic measure as well as a form of "netiquette." Regular interactions between pseudonymous participants can help to build trust and so engender meaningful deliberation. Nonetheless, the problems of anonymity are substantial.

The cyberactivists considered these issues in a later discussion, instigated by a poster who proposed that online debates should be governed in future by commonly agreed rules of discourse. The following extract formed part of a long and detailed post, which included proposals to widen participation by encouraging more women to join, and also to pursue links with Web sites of similar organizations such as Friends of the Earth. Responses were enthusiastic. It can be seen as an example of a virtual community tentatively beginning to think and act self-referentially:

Pairing up for dialogue is the way humans actually resolve disputes and come to recognize the limits of their own cognition. When two people (or even two bots [*sic*] posting propaganda) start a dialogue, we as a community should RESOLVE to provide supporting evidence or allay third party concerns or whatever we can do to ENSURE THAT THE DIALOGUE CONTINUES, especially if the two parties are overtly hostile. In other words, RESPECT DIALOGUE AS THE MOST BASIC FORM OF COMMUNICATION, not monologue. There are too many monologues here. Let's find specific ways to encourage pairs to debate visibly, e.g. threads are perhaps closed to all but two parties (each of which might have more than one poster) like a formal debate or which have two moderators, each with a overtly hostile view to the other. Let's focus on what we strongly disagree on. And get it worked out in detail if we can. Let's not allow arbitrary third parties to always pull the thread off track or make it hard to reach a conclusion. Let's find ways to support dialogue on this Greenpeace. org CAC board. (Cybercenter Archives, a: original emphasis)

In a short space of time, the Greenpeace Cybercenter has made impressive progress. The cyberactivists are actively exploring creative ways to improve their efficacy as an advocacy network, including ways to enhance the deliberative quality of their online forums. The Internet has served the needs of the participants in various ways according to their personal circumstances, but has evidently been invaluable for facilitating transborder political mobilization. These thoughtful contributions are indicative that shared interests can engender mutual feelings of affinity even when communication is solely computer-mediated.

6.7 Conclusion

Global civil society is the source of the most exciting applications of the Internet. As Castells observes, "it is in the realm of symbolic politics, and in the development of issue-oriented mobilizations by groups and individuals outside the mainstream political system that new electronic communication may have the most dramatic effects" (Castells 1997: 352). The vignettes of transnational activism discussed in this chapter present an intriguing picture of virtual networking. It represents only a small sample of the vast array and huge diversity of computer-mediated social movements. Evidently, counterhegemonic discourse is flourishing online. Through ICT, activists and NGOs are challenging entrenched power relations, publicizing the political claims of the disenfranchised, and raising the profile of marginalized social issues. The Internet is an essential vehicle for raising public consciousness about causes that are ignored by corporate media, and permits views to be expressed that are stifled elsewhere (Hill and Hughes, 1998). It has vastly expanded the opportunities for contact between like-minded individuals by transcending spatial-temporal barriers that obstruct physical interaction.

Progressive politics are well represented on the Internet. Nevertheless, as Warf and Grimes caution, "Counterhegemonic uses are not an electronic monopoly of the political left" (Warf and Grimes, 1997: 269). It is important to recall that civil society cannot be regarded as an intrinsic "force for good," and that bigotry, prejudice, and hatred equally find expression online. Yet reactionary movements can be adequately critiqued with reference to the norms of publicity if dialogue is premised on arbitrary claims about the exclusion of others. The benefit of public sphere theory is that it provides a method of normative evaluation to distinguish between progressive and repressive elements in civil society. Hence, the emancipatory potential of the above case studies can be suitably gauged if each is conceptualized within a public sphere frame of reference. I have suggested that the emergence of transnational networks of mutual affinity is defined by three criteria. Let me summarize the findings of this chapter by evaluating the case studies against each of these requirements.

The first criterion is *mutual affinity*. The case studies demonstrate quite clearly that mutual affinity is not dependent on territorial state-forms or a culturally specific social setting. Feelings of mutual affinity can arise instead in a virtual environment as a result of shared identities and interests. The strong solidaristic impulses of the international women's movement have led to campaigns to enhance female Internet access and usage. The participants of sites such as Cybergrrl, the Beijing discussion forum, and Greenpeace Cybercenter have testified to the attachment they feel toward their virtual networks, with evident sincerity. The Zapatista movement has been founded on broad-based protest at the inequities of neoliberalism, resulting in a surprisingly durable transnational coalition. What unites these case studies are frequent allusions to "community," and sentiments of fraternity and fellowship. This is indicative of a common identification with the social imaginary of a "public," which is a prerequisite for critical publicity. Of course, physical interaction maintains its value as a means of cultivating affinity. For example, the progress of virtual networking in the women's movement was expedited through links made at the Beijing Conference, and the Zapatista movement now holds annual global gatherings. Indeed, some Greenpeace activists express frustration that their involvement in the movement is restricted to virtual interaction. But face-to-face contact cannot be regarded as a necessary or sufficient condition for mutual affinity, merely another basis from which feelings of affinity can arise or be enhanced.

Second, public sphere interlocutors should exhibit a normative commitment to rational-critical discourse. So can a common endorsement of the *norms of publicity* be perceived amongst virtual networks? In each case study, there is evidence that participants engage in intelligible, reasoned debate, and are prepared to be publicly accountable for their opinions. The prime motivation behind each of these movements is to contact, engage, and converse with others. Members also actively pursue ways to increase inclusiveness. Examples include the women's groups that run Internet training workshops in the global South; the alternative media products that the Zapatistas produce for the disconnected rural poor; and the Greenpeace debates about widening participation. It cannot be denied that virtual networks are dominated by individuals and NGOs from the developed world. But stark statistics on the global digital divide may lead one to presuppose that virtual networks are less diverse than is suggested by the vibrant picture painted here. Who would expect the world's first "postmodern revolution" to be situated in rural Chiapas, or assume the maintenance of an Internet presence by RAWA during the years of Taliban government? These virtual networks were partly sustained by civil society in the global North, but the mainspring of creativity was sourced with the indigenous peoples. Hence, the Internet can foster inclusive dialogue by facilitating interconnections between different strata in world society.

However, as Bohman has noted, inclusivity is threatened by the inherent tendency of ICT to produce fragmented audiences. Sunstein (2001) has criticized discussion forums as resembling "echo chambers," attracting segmented audiences of like-minded persons, where accepted standpoints are unlikely to be subject to effective scrutiny. Consequently, forums may reinforce and shield people's worldviews from critical challenge. Such exclusivity and insularity is antithetical to a well-functioning public sphere. Previously, I have argued that some fragmentation occurs in any complex population, not least in the transnational realm. Segmented sites of discourse have also historically proved to be the sources of significant social transformation—the women's suffrage movement is a good example. Thus, segmented groups can still endorse the norms of publicity if discourse is sufficiently "public" in character. In other words, groups should seek discursive engagement with the outside world, and attempt to expand mainstream discourses. The case studies provide plentiful examples of this (most explicitly with the Zapatista's global consultation on their policy proposals).

The problems of social fragmentation are exacerbated when members use virtual networks in an instrumental way. The Greenpeace case study reveals that participation in the network can be self-interested and tactical. Perhaps this adds weight to the skeptical charge that the digital Trojan horse appears to be a social good but actually corrodes civic values by cultivating a culture of disengaged individualism. Further, anonymity can encourage authors to escape accountability for their actions and opinions, which can degrade trust and distort publicity. Virtual networks can be also subject to forms of publicity manipulation that are more commonly associated with mass media. Consider the way in which the political message of the Zapatista movement is excessively personified by Subcomandante Marcos. Overall, it is impossible to generalize about the deliberative quality of virtual networks because all are anchored in different socio-institutional contexts. As Downey and Fenton observe, "[I]nternet use is contributing simultaneously to new forms of social solidarity and fragmentation" (Downey and Fenton, 2003: 199). I have suggested certain ways in which the worst problems can be mitigated, but the difficulties are significant. Clearly, it is even harder to aspire to the norms of publicity in the transnational realm than it is domestically. But, importantly, there are signs that *it is not impossible*. For instance, the Greenpeace conversation about dialogic ground-rules represents a reflexive framing of public discourse, indicating critical-revolutionary potential.

The third criterion is *political efficacy*. Each of these case studies demonstrates that virtual networks can be dynamic enough to have wider political effects. The relationships that exist between social movements and sites of political authority vary a great deal. There is potential to build stronger links between decentered governance and deliberative networks. For instance, the Beijing conference represents an interesting point of engagement that could be further developed. Nonetheless, transformative effects on mainstream discourse and global governance are already manifesting. Campaign successes have been plentiful. Issue-areas have been brought to wider public attention. Decision-makers have also been pressurized to recognize the political claims of the marginalized. In particular, subordinate groups are empowered by the capacity of the Internet to disproportionately magnify the impact of localized initiatives. Each case study contained examples of the use of ICT to "globalize" resistance to local injustices, thereby demonstrating a dual configuration of political space.

These developments challenge traditional notions of bounded political community. The Zapatistas explicitly endorse this challenge by invoking discourse about the legitimacy and authority of the Mexican state, capitalism, and the international system (Morton, 2002). Cleaver attributes the recent growth of counterhegemonic networks as the "Zapatista Effect," which has resonated through social movements around the world, through the "impetus given to the active rejection of current policies, to the rethinking of the institutions and functioning of democracy and to the alternatives of the status quo" (Cleaver, 1998a: 622). Potentialities lie within these discourses for emancipatory transformation of world order.

In conclusion, I maintain that it is possible to identify transnational networks of mutual affinity. Notwithstanding some difficulties, the criteria proposed have largely been met. However, it is evident that the case studies differ substantially in terms of their internal structure and their relationship with sites of political authority. With regard to structure, the loose coalition that constitutes the Zapatista movement contrasts with the formal hierarchical organization of Greenpeace. With regard to governance, the Zapatistas have limited contact with formal political institutions (via sporadic and often hostile negotiations with the Mexican government), whereas some women's groups have established a largely constructive relationship with institutions such as the UN. The differences between different transnational networks require further theorization and research. This is a topic to which I shall return in the concluding chapter (section 7.2).

CHAPTER 7

Conclusion

The norms of publicity are fundamental for democratic theory, but have been perceived as having little application in the international realm. It is conventionally presumed that the international and the domestic are discrete domains, and that critical publicity is the exclusive preserve of the latter. The dichotomy is problematic in the current juncture, where peoples and societies are increasingly enmeshed in a web of social, political, and economic interconnections. Moreover, the global diffusion of ICT has generated a huge surge in cross-border dialogue and a concomitant interest in the normative potential of electronic transnational deliberation. IR has been slow to engage with public sphere theory, and the literature has only tentatively flirted with communication issues. Typically, analysis of ICT is shoehorned into a conventional theoretical (realist/liberal) framework. But what is needed—and what is often missing—is critical reflection about the challenges posed by the sociopolitical implications of ICT to the explanatory power of theoretical orthodoxies.

These omissions often reflect the shortcomings of conventional theory, which is characterized by a lack of reflexivity and tends to reify the status quo (Wyn Jones, 2001). In contrast, international critical theory is differentiated by an emancipatory orientation, and offers a powerful critique of global relations of power and domination. Yet even within critical IR, analysis of ICT is in short supply. Accordingly, key critical theorists have tended to undertheorize communicative aspects of emancipatory change. For example, Cox describes the multiple oppressions of hegemony in detail; but the deliberative mechanisms whereby these oppressions can be transcended are only sketched in brief (Comor, 1994). Public sphere theory is an ideal means of extending this direction of thought.

Indeed a number of scholars across a range of social science disciplines are conceiving of "globalized" publics (e.g., Bohman, 1997;

Dryzek, 1990; Hill and Montag, 2000). In the past few years, this debate has begun to cross over into IR, which is most welcome (e.g., Lynch, 1999, 2000; Mitzen, 2005). Public sphere theory introduces a useful new vocabulary to the discipline. However, a key problem of the IR/public sphere literature is also apparent in the wider debate on the "globalization of the public sphere": that is, the assumption that extraterritorial public spheres are actually existing institutions (e.g., Brunkhorst, 2002; Calhoun, 2003; Dryzek, 2000; Wilkin, 2001). Few theorists systematically investigate the institutional prerequisites for the extension of publicity. There is a danger that versions of the "extraterritorial public sphere" are entering the lexicon as common-sensical terms of reference without being fully grounded in theory or evidence. The rise in communication traffic certainly conveys a super-ficial impression of denationalized publics. But increases in the fre-quency and intensity of communication flow does not axiomatically equate to transnational public spheres.

Translating the norms of publicity to the international realm demands careful theorization, as the principles were originally devel-oped to apply in a domestic context. The norms center on the assertion that political authority is primarily legitimated through public opinion. Thus it is vital that deliberation within a political public sphere is linked with a mode of governance. As Nancy Fraser argues, this connection must be preserved or else the idea of a public sphere is depleted of its critical force and radical potential (Fraser, 2005). Public sphere theory only has analytical purchase for global politics if it is reconciled with its normative heritage. This inquiry is an attempt at such rapprochement. What follows is a brief recap of the main theoretical argument, final reflections on the findings, and some suggestions for future research.

7.1   A Review of the Main Thesis

This investigation is anchored on an appraisal of *STPS* by Jürgen Habermas, the locus classicus of public sphere theory. Habermas' account of the rise and decline of the early-modern bourgeois public was motivated by a normative interest in rationalizing state power. He claimed that there were historical indications that under certain social conditions, public opinion could effectively inform the organization of the polity. Thematic echoes from *STPS* pervade this inquiry, which is underpinned by a normative interest in rationalizing global gover-nance. The central question here is whether the potential exists in the present for international citizens to negotiate the organization of global society by using their capacities for reason and debate.

The public sphere has been traditionally conceptualized as commensurate with the nation-state. A number of implicit assumptions sustain this narrative. Among other things, the model of a territorially based public presupposes a national media and state sovereignty. This is unproblematic within the Habermasian rendition, which described a "transitional embodiment of the public sphere," located in eighteenth-century England (Habermas, 1992a: 467). However, the transformations associated with globalization impel a reassessment of these background presumptions (Held et al., 1999). The rise of ICT and transnational civil society has radicalized conventional notions of social and political space. Further, the growth of multilayered global governance suggests that state authority is becoming increasingly compromised (Ashley, 1988; Camilleri and Falk, 1992; Campbell, 1993; Walker, 1991, 1993; Weber, 1992). These challenges destabilize the institutional architecture of the domestic public sphere, but they might also lay the foundations for future structural transformation beyond the state.

It is important to recall that Habermasian theory was designed as a means to critically evaluate the democratic deficiencies of the nation-state. This is a mission shared by many of Habermas' critics, who object that the bourgeois public was structured by multiple social exclusions. For instance, Habermas has been castigated for neglecting the contribution of contemporaneous counterpublics, involving women, the working classes, racial minorities, and so on (e.g., Fraser, 1992; Negt and Kluge, 1993; Warner, 1992, 2002). Feminist theorists have shown how the bourgeois public/private dichotomy excludes a raft of gender-specific issues from debate (Pateman, 1987; Sylvester, 1994). It has been argued elsewhere that this artificial division precludes discussion on economic injustices (which adversely affects the working classes) and issues relating to sexual preference (which often discriminates against nonheterosexuals) (Eley, 1992; Norton, 1992). It is now widely accepted that Habermas exaggerated the extent of access and participation in the bourgeois public sphere. The critics have each proposed modified versions of the public that are designed to embody greater social inclusivity and broader terms of discourse. These revisions are important correctives to critical thought on state democratization. But the critiques remain limited in scope because they share a normative agenda framed by the nation-state. Consequently, the validity of a state-centric perspective has not been fully reassessed. It is imperative to expand the terms of reference for public sphere inquiry, since transnational deliberative spaces exist in the present with the *potential* to manifest critical publicity.

I submit an alternative approach based on a reconstruction of Habermasian theory and a review of the extraterritorial public spheres literature. It rests on the following definition: *a transnational public sphere is a site of deliberation in which non-state actors reach understandings about issues of common concern according to the norms of publicity.* The definition is intentionally functional, as the social and cultural diversity of the transnational realm is mirrored by multiple sites of variegated discourse. Several requirements are stipulated by the normative ideal of a public sphere. Debate should be open and inclusive, and participants must be regarded as nominal equals. All should endeavor to make their views intelligible and publicly accountable. Deliberation needs to be based on the exchange of reasons oriented toward understanding, and the public opinion thereby generated should be addressed to a site of political authority. These requirements are summarized as the *norms of publicity.*

Three structural preconditions are necessary for the emergence of transnational public spheres: transborder communicative capacity, transformation in sites in political authority, and transnational networks of mutual affinity. The preconditions are ideals that cannot wholly be realized in practice. Instead, they serve as guides for critical evaluation. However, an approximation of the ideals may occur around certain issue-areas. Therefore, an environment suitable for the emergence of transnational public spheres depends on the synthesis of each precondition around an issue-area.

7.1.1 Precondition One: Transborder Communicative Capacity

Habermas' conception of a political public was premised on a national mass media that primarily chronicled the activities of government (e.g., Habermas, 1999: 73). In recent years, media technology has been transformed beyond what would be possible to conceive in the early-modern era. There has been an extraordinary proliferation of media channels that cater for a kaleidoscopic range of interests. The advent of Internet technology enables interactive and decentralized forms of communication over vast distances. Also, trends of conglomeration and commercialization have produced a relative denationalization of communicative infrastructure and an oligopolistic global media market. Our hyper-networked societies are saturated with global media to an unprecedented extent, and are struggling to adapt to the consequences.

These developments contain emancipatory potentialities, which cannot be fully realized for several reasons. The global conversation is circumscribed by huge disparities of ICT access throughout the world. In particular, inequalities between the global North and South (but also within these regions) are an intractable obstacle to truly inclusive international dialogue (Tehranian, 1999). The problem is severely compounded by corresponding inequalities in education and income, which tend to reflect ethnic, sexual and social discriminations (Main, 2001). In addition, publicity is manipulated by overly powerful corporate actors. For instance, broadcast media fail to reflect the rich diversity of opinion in society, as they are primarily motivated by the quest for profit and tend to disproportionately privilege hegemonic discourses. Computer software monopolies have also distorted publicity through the imposition of rating and filtering technologies. Moreover, repressive states have used ICT to enhance their capacities for domination and control. State surveillance and censorship policies have curtailed freedom of expression in many parts of the world.

Still, it cannot be denied that the unique qualities of the Internet represent a veritable cornucopia of possibilities for political agency across state borders. Global diffusion of ICT has made steady progress— sometimes with unexpected rapidity. There is a general trajectory toward wider access, but much of this growth has been concentrated among the North and global elites. It would thus be unwise to discount the substantial obstacles to the emergence of transnational public spheres in terms of transborder communicative capacity. Therefore, it may be conjectured that privileged sections of world society can successfully aspire to the first precondition of transnational publics.

7.1.2   *Precondition Two: Transformations in Sites of Political Authority*

Conventional public sphere theory frames the nation-state as the addressee of public dialogue (e.g., Habermas, 1999: 81). The model is based on an assumption that there is a close correlation between sovereignty, territory and autonomy. Yet today there is mounting evidence that political authority is being dispersed and assigned to a variety of different agencies that operate at several levels—local, regional, and international. It is possible to argue that state sovereignty is a concept that has never been actually realized, but nonetheless remains analytically useful. However, critical inquiry about future transformations is restricted if public sphere theory is exclusively framed in the context of a sovereign and autonomous nation-state. It is impossible to ignore the international dimension in contemporary public sphere theory, because there is highly persuasive evidence that there have been structural changes in the modalities of governance.

The role of the nation-state is transforming due to the globalization of politics, the growth of international law and the relative decline in national identity. These trends are producing a structure of international authority that render the traditional conception of anarchy less relevant to an understanding of international politics. Our interconnected world can more accurately be conceived as administered through a matrix of multilayered global governance. Global governance is a somewhat confusing web of different actors, regimes, and jurisdictions. However, it is not without important centers in the circuits of power, neither does it imply the retreat of the state. Indeed, some of the most influential nodes in this tangled network are nation-states, which retain considerable powers and exclusive competencies. Crucially, states have nominal sovereignty in terms of entitlement to rule, which is a strategically important tool of political negotiation. However, under global governance, state sovereignty has been compromised in terms of the autonomy to act independently and to deliver policy programs. Decision making and policy formation is increasingly shaped by—and open to input from—a wide range of different actors. These developments suggest that it is unwise to locate undivided authority in national governments, and furthermore, that sites of global governance could be alternative addressees of transnational publicity.

The discursive connection between modes of governance and the citizenry is integral to the critical function of public spheres. The relationship is hugely complicated in emergent transnational publics in comparison to the domestic structural counterparts. Global governance is a complex and fluid arrangement of public and private actors. The source of effectual power is often vague and lines of accountability are oblique. Nevertheless it is encouraging to note that in many instances surveyed herein, activists appear to have quite a sophisticated understanding of which institutions impact upon their lives, and pitch their protests accordingly. The overall assessment is that the second precondition of transnational public spheres is extant.

7.1.3 Precondition Three: Transnational Networks of Mutual Affinity

Deliberative norms require that interlocutors recognize the moral legitimacy of public opinion generated through democratic discourse.

These norms are dependent on a minimal sense of commonality between the interlocutors, which I term "mutual affinity." In the conventional rendition of the public sphere, national citizenship engenders mutual affinity, and hence provides the social foundations for the common endorsement of the norms of publicity. However, mutual affinity does not seem to be the exclusive by-product of national citizenship. The previous chapter provided a wealth of examples that attest to the sociopolitical significance of global civil society. Many transnational social movements rely on vibrant virtual networks to converse, to problem-solve, to achieve consensus, and to articulate common interests and identities. Highly motivated members engage in intense discursive and political activity, and build meaningful relationships of solidarity with distant others. Naturally, there is massive diversity in the political aims and the quality of interaction between different groups, but it is undoubtedly possible to cultivate delocalized social bonds through ICT. The testimonials from the case study activists affirm that virtual networks are not inimical to empathetic feeling. There are frequent references to an imagined virtual "community," or to feelings of fraternity, that suggest that there is an inter-subjective understanding of the network as a "public."

It is more difficult to evaluate whether there is a sufficient degree of mutual affinity to realize the norms of publicity, especially when communication is largely computer-mediated. For instance, the building of trust between interlocutors is frustrated by the anonymity of Internet communication; and further aggravated when intercourse is based on the instrumental interests of the participants. Moreover, the principle of inclusivity can be jeopardized if online forums foster social fragmentation. These constraints inhibit the development of normatively structured discourse, but they are surmountable. In fact, seemingly "fragmented" counterpublics can be important agents in promoting greater inclusiveness in broader civic life, as demonstrated by the Zapatista advocacy of global dialogue. Also, some virtual networks have nurtured ties of mutual affinity through reflexive and unrestricted dialogue, either by tacitly or explicitly endorsing deliberative norms. For example, the Greenpeace Cybercenter has produced suggestions for forum rules that have been generated by the participants themselves. The potential to manifest higher forms of publicity clearly resides in such deliberative spaces.

Virtual networks are also dynamic enough to be politically efficacious. It is possible to detect the transformative influence of transnational public dialogue on hegemonic discourses and the global governance framework. The social movements examined here have

192 GLOBAL COMMUNICATION AND PUBLIC SPHERES

managed to shape the political agenda at critical junctures, and in some instances, they secured notable campaign successes. This indicates that some networks are sophisticated enough to operate within a highly complex governance environment, targeting different site/s of political authority according to their competencies—whether this be international organizations, multilateral forums, private actors, or nation-state governments. The points of engagement between governance and citizens tend to be transitory and inchoate, but formal institutionalization could be incipient, as demonstrated by the Beijing conference. However, enough evidence exists of politically potent sites of critical publicity to suggest that, in certain instances, the third precondition of transnational publics can be met.

Thus, despite the difficulties outlined above, it can be argued that the institutional foundations exist for the emergence of transnational public spheres. The exacting nature of the structural prerequisites means that they will only be realized in specific circumstances. Spheres are therefore likely to be episodic, located around issue-areas where there is a favorable confluence of communicative capacity, sites of global governance and activist networks.

The structural transformations in the preconditions of public spheres that have occurred since the early-modern period are profound, representing nothing less than a fundamental reordering of conventional notions of spatiality. The "information age" has rendered the temporal-spatial boundaries of public spheres increasingly fluid. The basic conditions of possibility for publics do not exclusively coincide with state borders. They have now been supplemented by transnational counterparts. The parallels and contrasts between the two have a seductive symmetry and comparisons are irresistible. The domestic public is physically bounded by the borders of the nation-state, whereas virtual networks occupy deterritorialized cyberspace. The national citizenry are physically proximate, yet transnational social movements are diffused over distant locations. Coffee shops and print media were eighteenth-century public forums; the twenty-first century complements are "cyber-salons" and digital technologies. Finally, framing all of these transformations is the institution of the nation-state: once popularly thought to be sovereign, now popularly thought to be destabilized because of the rise in global governance. The picture is one of materiality versus virtuality, monolithic hierarchies versus flexible networks. But this is close to simplistic caricature. The preconditions of domestic public spheres have been complicated by globalizing trends, but not necessarily invalidated. Domestic publics and emergent transnational spheres are not mutually exclusive; neither do the categories have a zero-sum relationship. Publics are now potentially amorphous, but can still assume state-forms. Indeed, some of the case study examples suggest that virtual networks can help state-based counterpublics to achieve their political aims. In such circumstances, it becomes difficult to distinguish the transnational from the domestic. The global resides in the local.

Nonetheless, transnational public spheres are little more than embryonic, and it is evident that consolidation is hindered by significant constraints on further progress. Global communication is systematically distorted, global governance is democratically deficient, and it is difficult to generate mutual affinity among highly differentiated virtual audiences. Unless these issues are addressed, the early promise of transnational networks as an emancipatory force will be neutralized. Does this mean that the concept of transnational publics is redundant, of little explanatory value? Far from it. Transnational public sphere theory does not lay claim to a revival of the Athenian agora through ICT, its ambitions are much more moderate. The transnational public sphere is an ideal-type that permits normative critique of actually existing conditions, and provides a means to analyze how global interconnectivity opens up prospects for new forms of politics. As such, it offers a promising avenue for the future development of critical international theory. I would like to conclude with some suggestions as to what direction this research could take.

7.2 THE FUTURE PATH FOR TRANSNATIONAL PUBLIC SPHERE THEORY

The next step for transnational public sphere research is to investigate the properties of emergent publics and the qualities of transnational deliberation. Further, the positive or adverse influence of certain factors on transnational critical publicity should be considered. These factors can be grouped into three categories, each directly relating to the structural prerequisites.

With regard to *transborder communicative capacity*, scholarly attention must be maintained on the social exclusions imposed by the global "digital divide," and the corresponding impacts on the potential to actualize more cosmopolitan forms of publicity. Comparative research on the distribution and application of different forms of ICT is imperative, especially in relation to faster modes of broadband and wireless Internet. There is an obvious danger that continuing sophistication of this technology will reverse the benefits accrued by Internet diffusion by consolidating the gap between the "info-rich" and "info-poor." It is also essential to chart and critique global trends in media ownership; and to document the manifold ways in which governments and corporations conspire to restrict free and open dialogue. Finally, detailed ethnographic case studies of virtual networks would be a useful venture. This inquiry has interrogated the concept of transnational publics by examining fundamental issues of participation and freedom of expression; follow-up questions might now be asked about the *qualities* of communication. For example, how well is reason deployed in public debate, and how responsive is public opinion to rational argument? Habermas employed his theory of communicative rationality in the state-based concept of the public sphere, and scholars such as Lincoln Dahlberg (2001) have explored ways of modifying these for a virtual environment.

With regard to *transformations in sites of political authority*, international critical theory must continue to attend to the growing "democratic deficit" that accompanies political globalization, and explore the ways in which transnational public spheres could increase the public accountability of global governance. Multilateral institutions are popularly perceived as remote and inaccessible to the public—effectively symbolized by the spectacle of WTO and G8 meetings sealed off from demonstrators by barricades and riot police. The institutions are largely answerable to member states that are ostensibly involved in their governance, and only open to indirect public participation. We have seen how NGOs and activist networks are overtly bypassing the state and making representations at an international level. The EU has formally institutionalized these channels of access through the European Parliament, the only directly elected element of the institution. European citizenry often mobilize around certain issues, lobby EU representatives and attempt to garner wider publicity among the electorate (Habermas, 2001). Scholarly energies should be devoted to examining these developments in more depth. On a case study basis, it may be possible to gain a deeper appreciation of the ways in which emergent transnational publics and governing institutions are mutually transformed by their encounters.

Theoretical and empirical research will help to unpack the broad categorization of *transnational networks of mutual affinity*. An unfortunate consequence of an umbrella term is that this implies some sort of equivalence between the groups, which is categorically not the case. Some networks are temporary coalitions of interest; some are an offshoot of an established NGO. Some are loose associations; others are part of a highly structured organization. It must be emphasized that these movements are not unambiguous symbols of critical-revolutionary potential, as they are often not fully democratic, inclusive, or accountable. Securing the equitable involvement of the greatest number in global governance will only be achieved if egalitarian principles are extended to global civil society. Thus, more work is needed on the ways in which differentials in internal composition impact on effective political communication, for which a comparative case study approach would be apposite.

Furthermore, it would be insightful to analyze the varying critical functions that transnational public spheres are beginning to assume, and the promise that this represents for the future organization of the political process. In this regard, I would like to suggest a possible way forward with reference to Nancy Fraser's distinction between weak and strong publics.

Habermas conceptualizes the public as an association of private persons, autonomous from the state, without direct command over the exercise of political authority. Fraser labels such public spheres as "*weak publics*, publics whose deliberative practices consists exclusively in opinion formation and does not also encompass decision making" (Fraser, 1992: 134, original emphasis). She distinguishes this from sovereign national parliaments, that are "*strong publics*" that act as a "public sphere *within* the state" (ibid.). Strong publics blur the separation between civil society and the state in their twin capacities for deliberation and administration. Fraser sees this as a democratic advance, since public opinion can translate into authoritative action. She argues that the prospects for evolution of strong publics are signaled by the rise in self-managing institutions, such as workplaces or residential committees, which establish "sites of direct or quasi-direct democracy" (135). Fraser's perspective suggests an intriguing way to conceive of emergent transnational public spheres. Habermas adapts the model in his "two-track" model of democracy in *Between Facts and Norms*, and Brunkhorst also explores the distinction between weak and strong publics in global politics (Habermas, 1996: Chapter 7; Brunkhorst, 2002). For each theorist, the defining requirement for a strong public is some measure of direct influence over the legislative process.

In a recent article, James Bohman also uses the distinction between weak and strong publics to theorize transnational deliberation. He considers that at present, transnational publics are limited to opinion formation, and so they can be defined as "weak." However, he argues, "they may become 'strong publics' when they are able to exercise influence through institutionalized decision procedures with regularized opportunities for *ex ante* input" (Bohman, 2004: 148). This requires that international institutions "permit such access to influence distributively, across various domains and levels," as "strong publics may be required to seek more direct forms of deliberative influence given the dispersal of authority and the variety of its institutional locations" (ibid.). In the international realm, the potential addressees of publicly generated opinion are manifold. Therefore, Bohman suggests that strong publics could emerge in a variety of institutional settings, where "publics are capable of exerting political influence in real decision-making processes under certain institutional conditions" (152). He follows Habermas' lead by suggesting that the EU holds interesting dialogic potential (Habermas, 2001). Likewise, Chalmers argues for reform of the EU to enable an effective deliberative approach to European governance (Chalmers, 2003). Also, Koopmans and Erbe have examined the Europeanization of public spheres, distinguishing between types of publics that relate to different European competencies. For example, they argue that strong publics have the potential to emerge concerning supranational policy areas such as monetary politics and agriculture, whereas intergovernmental areas such as education and pensions produce weak publics (Koopmans and Erbe, 2004). It seems likely that interest in the deliberative potential of the EU will increase in future.

This inquiry defined transnational public spheres as composed of non-state actors, and did not presuppose a formal influence over decision making or policy: these can be understood as "weak publics" according to Fraser's typology. My purpose has been to systematically investigate the prima facie grounds for transnational public spheres, and so I have expressly avoided speculation on what forms these publics might take. Now the conceptual terrain has been mapped out, it is possible to indulge such speculation. Hence future research into the properties of weak publics would be welcome, as would inquiry into the notion of strong transnational publics. However, it must be cautioned that it would be erroneous to presume that strong publics are actually existing institutions. In the first instance, the structural preconditions for the formal institutionalization of public spheres must be established. Thereafter, questions need to be asked about the exercise of power by strong publics, the relationship between weak and strong transnational publics, and the accountability of strong publics to weak. As Cochran proposes, it should also examine how transnational public spheres "would be internally regulated and made democratically accountable in the way that states are assumed to be in modern international politics" (Cochran, 1999: 271).

To conclude, research on transnational public spheres in IR is merely in its infancy. Where the concept has been introduced to the discipline, it has generally been divorced from the normative interest that was once regarded as an indivisible element of deliberative democracy. It is hoped that in the first instance, this investigation will serve as a useful road map for those wanting to conduct future research, and that in the second instance, public sphere theory has been successfully reconnected to the values that underpin the critical-theoretical project. This book opened with some reflections about the paradoxical nature of globalization. New media perfectly exemplifies the Janus-faced character of our times. On the one hand, ICT contributes to deepening social exclusion and has frightening capacities for repression and surveillance. On the other hand, the expansion of transnational deliberation inheres with emancipatory potentialities. There is a continuing struggle about which of these tensions will triumph. Immanent progressive tendencies could be cultivated to address the legitimacy crisis of global governance. But if these tantalizing prospects are to be realized, decision-makers face a tremendous challenge of imagination and political will. More importantly, the members of embryonic publics must embrace this challenge. The way forward will rely on participants to adopt the standpoint of what Bohman terms the "generalized other," "the relevant critical perspective that opens up a future standpoint of the whole community" (Bohman, 2004: 153). This dynamism depends on dialogue being maintained with those who serve the role of generalized other and expose the limitations of public spheres. Once obstacles to emancipation are identified and critiqued, then hope arises that they can be overcome. Yet as Bohman acknowledges, this will be so "only if there are agents who make it so and transnational institutions whose ideals seek to realize a transnational public sphere as the basis for a realistic utopia...in a complexly interconnected world" (154). The paradoxes of globalization have produced an abstract tug-of-war between the forces of repression and progression, which are locked in an unremitting battle for supremacy. The outcome is not predetermined; it is the responsibility of concerned citizens to enjoin the fight.

Notes

Chapter 1   Introducing Transnational Public Spheres to International Relations

1. An alternative version of the thesis herein has been published in the journal *Globalizations* (Crack, 2007).
2. I use the term "international relations" in lowercase letters to refer to the practice of international politics, law, and diplomacy. I use the capitalized variant, "International Relations" or the abbreviation "IR" to refer to the academic discipline.

Chapter 2   Reconstructing Habermasian Public Sphere Theory

1. Adorno analyzed the industrial production of cultural goods in capitalist society and identified a trend toward the production of culture as commodity (Adorno, 1991). He argued that this led to deterioration in the philosophical role of culture. Instead, mediated culture incorporated the proletariat into the structure of capitalism by limiting working-class ambitions to political and economic goals that could be met within the existing system. This discouraged the development of revolutionary consciousness. Adorno argued that this "culture industry" was therefore beneficial to the interests of the ruling classes. Expressed by Adorno and Horkheimer (1979) in its most pessimistic form, this critique suggests that mass culture has effectively extinguished opportunities to mobilize and conduct meaningful oppositional activity. Habermas has since conceded that

   > the strong influence of Adorno's theory of mass culture is not difficult to discern [in *Structural Transformation*]...At the time, I was too pessimistic about the resisting power and above all the critical potential of a pluralistic, internally much differentiated mass public. (Habermas, 1992a: 438)

2. This argument is heavily indebted to Nancy Fraser (2005). In addition to the points I make here, Fraser also claims that classical public sphere

theory presupposes the existence of a Westphalian–national economy subject to state regulation, a national language, and a national literary culture.

CHAPTER 3   CONTENDING THEORIES OF TRANSNATIONAL PUBLIC SPHERES: PROPOSITIONS FOR AN ALTERNATIVE ANALYTICAL FRAMEWORK

1. This is part of Dahlgren's wider conception of "civic cultures" underpinning extraterritorial public discourse (e.g., Dahlgren 2002, 2003, 2005).

CHAPTER 4   THE INFORMATION AGE: TRANSBORDER COMMUNICATIVE CAPACITY

1. "New media" refers to computer-mediated communications and to other technologies that have been digitally converged. I use the term "global media" to refer to new media as well as "older" forms of mass media, such as newspapers. However, it is important to note that this distinction is ambiguous in the context of digital networks (Axford and Huggins, 2001: vii).
2. Cyber-Rights and Cyber-Liberties is at http://www.cyber-rights.org/ (accessed May 1, 2007). The Electronic Frontier Foundation is at http://www.eff.org/Censorship/ (accessed May 1, 2007).
3. Also see the Amnesty International campaign against state censorship, Irrepressible.Info, at http://irrepressible.info/ (accessed May 1, 2007).
4. Information on the ITU "Connect the World" initiative is at http://www.itu.int/partners/index.html (accessed May 1, 2007). Stockholm Challenge information at http://www.stockholmchallenge.se (accessed May 1, 2007). Global Junior Challenge information at http://www.gjc.it/2006/en/index.php (accessed May 1, 2007).
5. Online Newspapers is at http://www.onlinenewspapers.com (accessed May 1, 2007).
   World Newspapers Online is at www.actualidad.com (accessed May 1, 2007).

CHAPTER 6   GLOBAL CIVIL SOCIETY: TRANSNATIONAL NETWORKS OF MUTUAL AFFINITY

1. For more information, see http://www.rawa.org/index.php (accessed May 1, 2007).

2. For more information, refer to http://www.modemmujer.org/ (accessed May 1, 2007).
3. Information on Womenwatch is at http://www.un.org/womenwatch/ (accessed May 1, 2007).
4. This quote is taken from the homepage of WomenAction, where more information on the project can be accessed. See http://www. womenaction.org/ (accessed May 1, 2007).
5. For example, see the official EZLN site, Enlace Zapatista at http:// enlacezapatista.ezln.org.mx/ (accessed May 1, 2007). The Chiapas Indymedia site is at http://chiapas.indymedia.org/ (accessed May 1, 2007). The ZNet Chiapas Watch page is at http://www.zmag.org/ chiapas1/index.htm (accessed May 1, 2007). A Zapatista Discussion Group is at http://www.zapatistas.org/.
6. Accion Zapatista de Austin can be found at http://www.eco.utexas. edu/~hmcleave/chiapas95.html (accessed May 1, 2007).
7. Information on the Irish Mexico Group is at http://flag.blackened. net/revolt/mexico/img/irimex.html (accessed May 1, 2007).
8. Radio Insurgente can be accessed at http://www.radioinsurgente. org/ (accessed May 1, 2007).
9. See the Accion Zapatista de Austin site, as detailed above.
10. See http://www.greenpeace.org (accessed May 1, 2007).
11. The comments posted by supporters on the Cybercenter Web site, including the extracts published in this book, do not necessarily reflect the views of Greenpeace. The archives of the Cyberactivist Center can be accessed at http://activism.greenpeace.org/cybercentre/ (accessed May 1, 2007). It has recently been superseded by a new discussion forum, LouderThanWords. See http://forum.greenpeace. org/int/ (accessed May 1, 2007).

BIBLIOGRAPHY

ACLU. USA Patriot Act: Section 215. American Civil Liberties Union. Available from http://action.aclu.org/reformthepatriotact/215.html (accessed May 1, 2007).

Adams, P. C. 1996. Protest and the Scale Politics of Telecommunication. *Political Geography* 15 (5): 419–441.

Adorno, T. 1991. *The Cultural Industry: Selected Essays on Mass Culture.* London: Routledge.

Adorno, T., and Horkheimer, M. 1979. *Dialectic of Enlightenment.* London: Verso.

Ahmed, I. 2002. Globalisation and Human Rights in Pakistan. *International Journal of Punjab Studies* 9 (1): 57–89.

Albrow, M. 1996. *The Global Age.* Cambridge: Polity.

Alegre, N. 2003. From the Pen to the Keyboard in Mexico. International Institute for Communication and Development. Available from http://www.iicd.org/photos/iconnect/Stories/Story.import5169 (accessed May 1, 2007).

Alger, C. 2002. The Emerging Roles of NGOs in the UN System: From Article 71 to a People's Millennium Assembly. *Global Governance* 8 (1): 93–117.

Alter, K. 1996. The European Court's Political Power. *West European Politics* 19 (3): 458–487.

———. 1998. "Who Are the Masters of the Treaty?" European Governments and the European Court of Justice. *International Organization* 52 (1): 121–147.

Amnesty International. 2000. *Annual Report.* Available from http://web.amnesty.org/report2000 (accessed May 1, 2007).

———. 2004. People's Republic of China. Controls Tighten as Internet Activism Grows. Available from http://web.amnesty.org/library/Index/ENGASA170012004 (accessed May 1, 2007).

———. 2006. *Undermining Freedom of Expression in China. The Role of Yahoo!, Microsoft and Google.* London: Amnesty International.

Amoore, L., and P. Langley. 2004. Ambiguities of Global Civil Society. *Review of International Studies* 30 (1): 89–110.

Anderson, B. 1991. *Imagined Communities.* London: Verso.

Anheier, H., M. Glasius, and M. Kaldor. 2001. Introducing Global Civil Society. In *Global Civil Society 2001.* Ed. H. Anheier, M. Glasius, and M. Kaldor, 3–22. Oxford: Oxford University Press.

Antitrust Division. *United States v. Microsoft*. U.S. Department of Justice. Available from http://www.usdoj.gov/atr/cases/ms_index.htm (accessed May 1, 2007).

Antrobus, P. 2004. *The Global Women's Movement: Origins, Issues and Strategies*. London: Zed Books.

APC. Using the Internet Strategically. Association for Progressive Communications. Available from http://www.apc.org/english/capacity/strategy/examples_90s.shtml#index (accessed May 1, 2007).

Arabic Network for Human Rights Information. 2006. Syria. *Implacable Adversaries: Arab Governments and the Internet*. Available from http://www.openarab.net/en/reports/net2006/syria.shtml (accessed May 1, 2007).

Aravamudan, S. 1999. *Tropicopolitans: Colonialism and Agency, 1688–1804*. Durham: Duke University Press.

Arendt, H. 1958. *The Human Condition*. Chicago: University of Chicago Press.

Aronson, J. 2005. The Communications and Internet Revolution. In *The Globalization of World Politics*. Ed. J. Baylis and S. Smith, 540–558. Oxford: Oxford University Press.

Arquilla, J., and D. Ronfeldt. 1996. *The Advent of Netwar*. Santa Monica: RAND.

Ashley, R. K. 1988. Untying the Sovereign State: A Double Reading of the Anarchy Problematique. *Millennium* 17 (2): 227–262.

Ashley, R. K., and R. B. J. Walker. 1990. Reading Dissidence/Writing the Discipline: Crisis and the Question of Sovereignty in International Studies. *International Studies Quarterly* 34 (3): 367–412.

Autonomedia. 1994. *Zapatistas! Documents of the New Mexican Revolution*. Latin American Network Information Center. Available from http://lanic.utexas.edu/project/Zapatistas/index.html (accessed May 1, 2007).

Axford, B. 2001. The Transformation of Politics or Anti-politics? In *New Media and Politics*. Ed. B. Axford and R. Huggins, 1–29. London: Sage.

Axford, B., and R. Huggins, eds. 2001. *New Media and Politics*. London: Sage.

Bagdikian, B. 2004. *The New Media Monopoly*. 7th ed. Boston, MA: Beacon Press.

Baudrillard, J. 1995. *The Gulf War Did Not Take Place*. Bloomington: Indiana University Press.

Baynes, K. 2001. Deliberative Politics, the Public Sphere and Global Democracy. In *Critical Theory and World Politics*. Ed. R. Wyn Jones, 161–170. London: Lynne Rienner.

BBC. 2000. Free Speech Advocates Criticise AOL Merger. BBC News: Business, January 12. Available from http://news.bbc.co.uk/1/hi/business/600713.stm (accessed May 1, 2007).

———. 2006. *Annual Report and Accounts 2005/6*. London: BBC. Available from http://news.bbc.co.uk/1/hi/business/6408391.stm (accessed May 1, 2007).

———. 2007. Microsoft Warned of More EU Fines. *BBC News: Business*, March 1. Available from http://news.bbc.co.uk/1/hi/business/6408391. stm (accessed May 1, 2007).

Benard, C. 2002. *Veiled Courage: Inside the Afghan Women's Resistance.* New York: Broadway Books.

Bender, T. 1978. *Community and Social Change in America.* New Brunswick, NJ: Rutgers University Press.

Benhabib, S. 1992. Models of Public Space: Hannah Arendt, the Liberal Tradition, and Jürgen Habermas. In *Habermas and the Public Sphere.* Ed. C. Calhoun, 73–97. Cambridge, MA: MIT Press.

———. 1995. The Pariah and Her Shadow: Hannah Arendt's Biography of Rahel Varnhagen. *Political Theory* 23 (4): 5–24.

———. 2002. *The Claims of Culture: Equality and Diversity in the Global Era.* Princeton: Princeton University Press.

Blaug, R. 1994. Habermas's Treatment for Relativism. *Politics* 14 (2): 51–57.

Bohman, J. 1997. The Public Spheres of the World Citizen. In *Perpetual Peace: Essays on Kant's Cosmopolitan Ideal.* Ed. J. Bohman and M. Lutz-Bachmann, 179–200. Cambridge, MA: MIT Press.

———. 1998. The Globalization of the Public Sphere. *Philosophy and Social Criticism* 24 (2/3): 199–216.

———. 2004. Expanding Dialogue: The Internet, the Public Sphere and Prospects for Transnational Democracy. In *After Habermas: New Perspectives on the Public Sphere.* Ed. N. Crossley and J. M. Roberts, 131–155. Oxford: Blackwell.

Boli, J., and T. A. Loya. 1999. National Participation in World-Polity Organizations. In *Constructing World Culture: International Nongovernmental Organizations since 1875.* Ed. J. Boli and G. M. Thomas, 50–77. Stanford: Stanford University Press.

Booth, K. 1995. Human Wrongs and International Relations. *International Affairs* 71 (1): 103–126.

Boyd-Barrett, O. 1995. Conceptualizing the "Public Sphere." In *Approaches to Media: A Reader.* Ed. O. Boyd-Barrett and C. Newbold, 200–234. New York: St. Martin's Press.

Bray, J. 2000. Tibet, Democracy and the Internet Bazaar. *Democratization* 7 (1): 157–173.

Brodsky, A. 2003. *With All Our Strength: The Revolutionary Association of the Women of Afghanistan.* New York: Routledge.

Brown, C. 1992. *International Relations Theory: New Normative Approaches.* London: Harvester Wheatsheaf.

Brown, M. 2005. Abandoning the News. *Carnegie Reporter* 3 (2). Available from http://www.carnegie.org/reporter/10/news/index.html (accessed May 1, 2007).

Brown, P. 2000. Europe Votes for a Stop in Nuclear Waste Reprocessing. *The Guardian*, June 30. Available from http://www.guardian.co.uk/uk_news/story/0,,338114,00.html (accessed May 1, 2007).

Brunkhorst, H. 2002. Globalising Democracy Without a State: Weak Public, Strong Public, Global Constitutionalism. *Millennium* 31 (3): 675–690.

Buck-Morss, S. 2002. A Global Public Sphere? *Situation Analysis* (1): 1–10.

Bull, H. 1966. International Theory: The Case for a Classical Approach. *World Politics* 18 (1): 361–377.

Burbach, R. 1994. Roots of the Postmodern Rebellion in Chiapas. *New Left Review* (May/June): 113–125.

Burchill, S. 1996. Realism and Neo-realism. In *Theories of International Relations.* Ed. S. Burchill and A. Linklater, 67–92. London: Macmillan.

Burnett, R. 1995. *Cultures of Vision: Images, Media and the Imaginary.* Bloomington: Indiana University Press.

Cable, V. 1999. *Globalization and Global Governance.* London: Royal Institute of International Affairs.

Calhoun, C. 1989. Tiananmen, Television and the Public Sphere: Internationalization of Culture and the Beijing Spring of 1989. *Public Culture* 2 (1): 54–71.

———, ed. 1992. *Habermas and the Public Sphere.* Cambridge, MA: MIT Press.

———. 1995. *Critical Social Theory: Culture, History and the Challenge of Difference.* Oxford: Blackwell.

———. 2003. Information Technology and the International Public Sphere. In *Shaping the Network Society: The New Role of Civil Society in Cyberspace.* Ed. D. Schuler and P. Day, 229–251. Cambridge, MA: MIT Press.

Camilleri, J. A., and J. Falk. 1992. *The End of Sovereignty?* Aldershot: Edward Elgar.

Campbell, D. 1993. *Sovereignty, Ethics and the Narratives of the Gulf War.* Boulder, CO: Lynne Rienner.

———. 2001. Mexico's Masked Man Strides into the Capital. *The Guardian.* March 12: 13.

Carr, E. H. 1939. *The Twenty Years' Crisis, 1919–1939.* London: Macmillan.

Cassese, A. 1988. *Violence and Law in the Modern Age.* Cambridge: Polity.

Castells, M. 1997. *The Information Age: Economy, Society and Culture: Vol. II. The Power of Identity.* Oxford: Blackwell.

———. 1998. *The Power of Identity.* Oxford: Basil Blackwell.

Cerny, P. G. 1995. Globalization and the Changing Logic of Collective Action. *International Organization* 49 (4): 595–625.

———. 1996. What Next for the State? In *Globalization: Theory and Practice.* Ed. E. Kofman and G. Youngs, 123–137. London: Pinter.

———. 1999. Globalization, Governance and Complexity. In *Globalization and Governance.* Ed. A. Prakash and J. A. Hart, 188–212. London: Routledge.

Cevallos, D. 1998. Politics-Mexico: Government Lost and Defeated in Cyberspace. Inter Press Service News Agency. Available from http://ipsnews.net/latin.asp (accessed May 1, 2007).

Chadwick, A. 2006. *Internet Politics: States, Citizens and New Communication Technologies.* Oxford: Oxford University Press.

Chalmers, D. 2003. The Reconstitution of European Public Spheres. *European Law Journal* 9 (2): 127–189.

Charnowitz, S. 1997. Two Centuries of Participation: NGOs and International Governance. *Michigan Journal of International Law* 18 (1): 183–286.

Chatterjee, P., and M. Finger. 1994. *The Earth Brokers: Power, Politics and World Development.* London: Routledge.

Checkel, J. T. 1998. The Constructivist Turn in International Relations Theory. *World Politics* 50 (2): 324–348.

Chevaldonne, F. 1987. Globalisation and Orientalism: The Case of TV Serials. *Media, Culture & Society* 9 (1): 137–148.

Chin, A. 2005. Decoding *Microsoft*: A First Principles Approach. *Wake Forest Law Review* 40 (1): 1–157.

Chroust, P. 2000. Neo-Nazis and Taliban On-Line: Anti-modern Political Movements and Modern Media. *Democratization* 7 (1): 102–117.

Clark, I. 1997. *Globalization and Fragmentation: International Relations in the Twentieth Century.* Oxford: Oxford University Press.

Cleaver, H. J. 1998a. The Zapatista Effect: The Internet and the Rise of an Alternative Political Fabric. *Journal of International Affairs* 51 (2): 621–640.

———. 1998b. Zapatistas and the Electronic Fabric of Struggle. In *Zapatista! Reinventing Revolution in Mexico.* Ed. J. Holloway and E. Peláez, 81–103. London: Pluto Press.

———. 1999. Computer-Linked Social Movements and the Global Threat to Capitalism. Available from http://www.eco.utexas.edu/faculty/Cleaver/polnet.html (accessed May 2007).

Cobain, I. 2006. The Internet's Role. *The Guardian*, October 7. Available from http://business.guardian.co.uk/story/0,,1889778,00.html (accessed May 1, 2007).

Cochran, M. 1999. *Normative Theory in International Relations: A Pragmatic Approach.* Cambridge: Cambridge University Press.

CoE. 2003. Additional Protocol to the Convention on Cybercrime, concerning the Criminalisation of Acts of a Racist and Xenophobic Nature Committed through Computer Systems. Council of Europe. Available from http://conventions.coe.int/Treaty/en/Treaties/Html/189.htm (accessed May 1, 2007).

Cohen, E. S. 2001. Globalization and the Boundaries of the State: A Framework for Analyzing the Changing Practice of Sovereignty. *Governance* 14 (1): 75–97.

Cohen, R., and S. M. Rai, eds. 2000. *Global Social Movements.* London: Athlone Press.

Colás, A. 2002. *International Civil Society: Social Movements in World Politics.* Cambridge: Polity.

Comor, E. A., ed. 1994. *The Global Political Economy of Communication.* London: Macmillan.

Compaine, B. 2002. Global Media. *Foreign Policy* 133 (November–December): 20–28.

Compaine, B. 2005. The Media Monopoly Myth. How New Competition Is Expanding Our Sources of Information and Entertainment. New Millennium Research Council. Available from http://www. newmillenniumresearch. org/archive/Final_Compaine_Paper_050205.pdf (accessed May 1, 2007).

Cooke, M. 1994. *Language and Reason: A Study of Habermas's Pragmatics.* Cambridge, MA: MIT Press.

Cope, B., and M. Kalantzis. 2000. *Multiliteracies: Literacy Learning and the Design of the Social Future.* London: Routledge.

Corn-Revere, R. 2002. Caught in the Seamless Web: Does the Internet's Global Reach Justify Less Freedom of Speech? Washington, DC: CATO Institute.

Covell, A. 1999. *Digital Convergence: How the Merging of Computers, Communications and Multimedia Is Transforming Our Lives.* Newport, RI: Aegis.

Cox, R. W. 1981. Social Forces, States and World Order: Beyond International Relations Theory. *Millennium* 10 (2): 126–155.

———. 1983. Gramsci, Hegemony and International Relations: An Essay in Method. *Millennium* 12 (2): 162–175.

———. 1987. *Production, Power and World Order: Social Forces in the Making of History.* New York: Columbia University Press.

———. 1992. Towards a Post-hegemonic Conceptualization of World Order: Reflections on the Relevancy of Ibn Khaldun. In *Governance Without Government: Order and Change in World Politics.* Ed. J. N. Rosenau and E.-O. Czempiel, 132–159. Cambridge: Cambridge University Press.

———. 1999. Civil Society at the Turn of the Millennium: Prospects for an Alternative World Order. *Review of International Studies* 25 (1): 3–28.

Crack, A. M. 2004. The Structural Preconditions for the Emergence of Transnational Public Spheres. Diss., Southampton, UK: University of Southampton.

———. 2007. Transcending Borders? Reassessing Public Spheres in a Networked World. *Globalizations* 4 (3): 341–354.

Crossley, N., and J. M. Roberts, eds. 2004. *After Habermas: New Perspectives on the Public Sphere.* Oxford: Blackwell.

Croteau, D., and W. Hynes. 2005. *The Business of Media: Corporate Media and the Public Interest.* 2nd ed. London: Sage.

CSIA, 2006. Internet Security National Survey, no. 3. May 23. Cyber Security Industry Alliance. Available from https://www.csialliance.org/resources/publications/CSIA_Internet_Survey_May_2006.PDF (accessed May 1, 2007).

Curran, J., ed. 1991. Rethinking the Media as a Public Sphere. In *Communication and Citizenship.* Ed. P. Dahlgren and C. Sparks, 38–42. London: Routledge.

Cybercenter Archives, a. Getting Deeper into Activism (discussion thread). Greenpeace Cyberactivist Community. Available from http://activism. greenpeace.org/cybercentre/gpi/Cybercentre/1043612859/10436128 59.html (accessed May 1, 2007).

———, b. Haunted Bhopal (discussion thread). Greenpeace Cyberactivst Community. Available from http://activism.greenpeace.org/cybercentre/gpi/Toxics/1007061842/1007061842.html (accessed May 1, 2007).

———, c. One Year On (discussion thread). Greenpeace Cyberactivist Community. Available from http://activism.greenpeace.org/cybercentre/gpi/Cybercentre/1006302573/1006302573.html (accessed May 1, 2007).

Dahlberg, L. 2001. The Internet and Democratic Discourse: Exploring the Prospects of Online Deliberative Forums Extending the Public Sphere. *Information, Communication & Society* 4 (4): 615–633.

Dahlgren, P. 2002. In Search of the Talkative Public: Media, Deliberative Democracy and Civic Culture. *Javnost - The Public* 9 (3): 1–21.

———. 2003. Reconfiguring Civic Culture in the New Media Milieu. In *Media and Political Style: Essays on Representation and Civic Culture.* Ed. J. Corner and D. Pels, 151–170. London: Sage.

———. 2005. The Internet, Public Spheres and Political Communication: Dispersion and Deliberation. *Political Communication* 22 (2): 147–162.

Dahlgren, P., and C. Sparks, eds. 1991. *Communication and Citizenship: Journalism and the Public Sphere in the New Media Age.* London: Routledge.

De Huerta, M. D., and N. Higgins. 1999. An Interview with Subcomandante Insurgente Marcos, Spokesperson and Military Commander of the Zapatista Uprising in Chiapas, Mexico. *International Affairs* 75 (2): 269–279.

Dean, J. 2001. Cybersalons and Civil Society: Rethinking the Public Sphere in Transnational Technoculture. *Public Culture* 13 (2): 243–262.

Deibert, R. J. 1997. *Parchment, Printing, and Hypermedia: Communication in World Order Transformation.* New York: Columbia University Press.

Deibert, R. J., and N. Villeneuve. 2004. Firewalls and Power: An Overview of Global State Censorship of the Internet. In *Human Rights in the Digital Age.* Ed. M. Klang and A. Murray, 111–124. London: GlassHouse.

della Porta, D., H. Kriesi, and D. Rucht, eds. 1999. *Social Movements in a Globalizing World.* Basingstoke, UK: Macmillan.

Der Derian, J., and J. Shapiro. 1988. *International/Intertextual Relations: Postmodern Readings in World Politics.* Lexington: Lexington Books.

Deutsch, K. W. 1957. Mass Communication and the Loss of Freedom in National Decision-Making: A Possible Research Approach to Interstate Conflict. *Journal of Conflict Resolution* 1 (2): 200–211.

———. 1963. *The Nerves of Government: Models of Political Communication and Control.* New York: Free Press.

———. 1966. *Nationalism and Social Communication.* Cambridge, MA: MIT Press.

Dewey, J. 1927. *The Public and Its Problems.* London: Allen & Unwin.

Dickenson, D. 1997. Counting Women In: Globalization, Democratization and the Women's Movement. In *The Transformation of Democracy?*

Globalization and Territorial Democracy. Ed. A. McGrew, 97–120. Cambridge: Polity.

Donnelly, J. 2000. *Realism and International Relations.* Cambridge: Cambridge University Press.

Douglas, S. J. 1995. *Where the Girls Are Growing Up Female with the Mass Media.* London: Penguin.

Doward, J. 2006a. How Korea Reached the Top of Its Game. In Special Report Section: The Broadband Revolution. *The Observer,* December 10: 6–7.

———. 2006b. We're Wired.... But Not Connected. In Special Report Section: The Broadband Revolution. *The Observer,* December 10: 1–2.

Downey, J., and N. Fenton. 2003. New Media, Counter Publicity and the Public Sphere. *New Media & Society* 5 (2): 185–202.

Doyle, M. 1986. Liberalism and World Politics. *American Political Science Review* 80 (4): 1151–1169.

Drainville, A. 1998. The Fetishism of Global Civil Society: Global Governance, Transnational Urbanism and Sustainable Capitalism in the World Economy. In *Transnationalism from Below.* Ed. M. P. Smith and L. E. Guarnizo, 35–63. London: Transaction Publishers.

Dryzek, J. S. 1990. *Discursive Democracy: Politics, Policy and Political Science.* Cambridge: Cambridge University Press.

———. 2000. *Deliberative Democracy and Beyond.* Oxford: Oxford University Press.

Eley, G. 1992. Nations, Publics and Political Cultures: Placing Habermas in the Nineteenth Century. In *Habermas and the Public Sphere.* Ed. C. Calhoun, 289–339. Cambridge, MA: MIT Press.

Elshtain, J. B. 1995. *Women and War.* Chicago: University of Chicago Press.

Enloe, C. 1990. *Bananas, Beaches and Bases: Making Feminist Sense of International Relations.* Berkeley: California University Press.

EPIC. 1997. Faulty Filters: How Content Filters Block Access to Kid-Friendly Information on the Internet. Electronic Privacy Information Center. Available from http://www2.epic.org/reports/filter-report.html (accessed May 1, 2007).

EU Commission. 2006. *Bridging the Broadband Gap.* European Union. Available from http://eur-lex.europa.eu/LexUriServ/site/en/com/2006/com2006_0129en01.pdf (accessed May 1, 2007).

EZLN. 1994a. *First Declaration from the Lacandon Jungle "Today We Say: Enough Is Enough!"* Ejército Zapatista de Liberación Nacional. Available from http://www.ezln.org/documentos/1994/199312xx.en.htm (accessed May 1, 2007).

———. 1994b. *Second Declaration from the Lacandon Jungle. "Today We Say: We Will Not Surrender!"* Ejército Zapatista de Liberación Nacional. Available from http://www.ezln.org/documentos/1994/19940610.en.htm (accessed May 1, 2007).

FAIR. 2007. Iraq and the Media: A Critical Timeline. Fairness and Accuracy in Reporting. Available from http://www.fair.org/index.php?page=3062 (accessed May 1, 2007).

Falk, R. 1998. Global Civil Society: Perspectives, Initiatives, Movements. *Oxford Development Studies* 16 (1): 99–110.

Fallows, D. 2005. *How Women and Men Use the Internet.* Pew Internet and American Life Project. Available from http://www.pewinternet.org/pdfs/PIP_Women_and_Men_online.pdf (accessed May 1, 2007).

Fassbender, B. 1998. The United Nations Charter as Constitution of the International Community. *Columbia Journal of Transnational Law* 36: 529–619.

Feldman, L. C. 2002. Redistribution, Recognition and the State: The Irreducibly Political Dimension of Injustice. *Political Theory* 30 (3): 410–440.

Ferdinand, P. 2000. The Internet, Democracy and Democratization. *Democratization* 7 (1): 1–17.

Financial Times. 2006. FT Global 500. Available from http://media.ft.com/cms/adb61f66-f7bf-11da-9481-0000779e2340,dwp_uuid=c9034b2c-f175-11da-940b-0000779e2340.pdf (accessed May 1, 2007).

Foster, D. 1997. Community and Identity in the Electronic Village. In *Internet Culture.* Ed. D. Porter, 23–37. London: Routledge.

Foster, M. 2006. Freedom and Friendships Formed in Cyberspace. In Special Report Section: The Broadband Revolution. *The Observer*, December 10: 7.

Fox, S. 2005. Digital Divisions. Pew Internet and American Life Project. Available from http://www.pewinternet.org/pdfs/PIP_Digital_Divisions_Oct_5_2005.pdf (accessed May 1, 2007).

Fraser, N. 1992. Rethinking the Public Sphere: A Contribution to the Critique of Actually Existing Democracy. In *Habermas and the Public Sphere.* Ed. C. Calhoun, 109–143. Cambridge, MA: MIT Press.

———. 1995. Recognition or Redistribution? A Critical Reading of Iris Young's Justice and the Politics of Difference. *Journal of Political Philosophy* 3 (2): 166–180.

———. 2005. Transnationalizing the Public Sphere. European Institute for Progressive Cultural Policies. Available from http://www.republicart.net/disc/publicum/fraser01_en.htm (accessed May 1, 2007).

Fraser, N., and A. Honneth. 2003. *Redistribution or Recognition? A Political-Philosophical Exchange.* New York: Verso.

Frederick, H. H. 1993. *Global Communication and International Relations.* London: Harcourt Brace College Publishers.

Froehling, O. 1997. The Cyberspace "War of Ink and Internet" in Chiapas, Mexico. *Geographical Review* 87 (2): 291–307.

G8. 2000. Okinawa Charter on the Global Information Society. University of Toronto G8 Information Centre. Available from http://www.g7.utoronto.ca/summit/2000okinawa/gis.htm (accessed May 1, 2007).

Gallaher, C., and O. Froehling. 2002. New World Warriors: "Nation" and "State" in the Politics of the Zapatista and US Patriot Movements. *Social and Cultural Geography* 3 (1): 81–102.

Gandy, O. H. J. 1988. The Political Economy of Communications Competence. In *The Political Economy of Communication*. Ed. V. Mosco and J. Wasko, 108–124. Madison, WI: University of Wisconsin Press.

Garnham, N. 1992. The Media and the Public Sphere. In *Habermas and the Public Sphere*. Ed. C. Calhoun, 359–376. Cambridge, MA: MIT Press.

George, J. 1994. *Discourses of Global Politics: A Critical (Re)Introduction to International Relations*. Boulder, CO: Lynne Rienner.

Giddens, A. 1990. *The Consequences of Modernity*. Cambridge: Polity.

———. 1991. *Modernity and Self-Identity*. Cambridge: Polity.

———. 1994. *Beyond Left and Right. The Future of Radical Politics*. Stanford, CA: Stanford University Press.

Gill, S. 1995. The Global Panopticon? The Neoliberal State, Economic Life, and Democratic Surveillance. *Alternatives* 20 (1): 1–49.

Gilroy, P. 1993. *The Black Atlantic: Modernity and Double Consciousness*. Cambridge, MA: Harvard University Press.

Gimmler, A. 2001. Deliberative Democracy, the Public Sphere and the Internet. *Philosophy and Social Criticism* 27 (4): 21–40.

Global Reach. *Global Internet Statistics (by Language)* Available from http://www.glreach.com/globstats/index.php3 (accessed May 1, 2007).

Glosserman, B. 1996. How Green Is My Cyberspace? *Japan Times Weekly International Edition* 37 (4): 15.

Goldstein, L. F., and Ban, C. 2005. The European Human Rights Regime as a Case Study in the Emergence of Global Governance. In *Contending Perspectives on Global Governance*. Ed. A. D. Ba and M. J. Hoffman, 154–177. London, Routledge.

Graham, J. 2006. RIAA Chief Says Illegal File-Sharing "Contained." *USA Today*, June 12. Available from http://www.usatoday.com/tech/products/services/2006-06-12-riaa_x.htm (accessed May 1, 2007).

Greenburg, M. H. 2003. A Return to Lilliput: The *Licra v. Yahoo!* Case and the Regulation of Online Content in the World Market. *Berkeley Technology Law Journal* 18 (4): 1191–1258.

Greenpeace, a. Cyberactivism Revolutionizes Greenpeace Campaigns. Greenpeace Archive. Available from http://archive.greenpeace.org/cyberstory/cyberactivism.htm (available May 1, 2007).

———, b. Greenpeace's E-mail, Internet and WWW History. Greenpeace Archive. Available from http://archive.greenpeace.org/history.shtml (accessed May 1, 2007).

Griffin, K. 2003. Economic Globalization and the Institutions of Global Governance. *Development and Change* 34 (5): 789–808.

Guidry, J. A., M. D. Kennedy, and M. N. Zald, eds. 2000. *Globalizations and Social Movements: Culture, Power, and the Transnational Public Sphere*. Ann Arbor, MI: University of Michigan Press.

Gunder Frank, A. 1980. *Crisis in the World Economy*. London: Heinemann.

Gunkel, D. J. 2003. Second Thoughts: Toward a Critique of the Digital Divide. *New Media & Society* 5 (4): 499–522.

Haacke, J. 1996. Theory and Praxis in International Relations: Habermas, Self-Reflection, Rational Argumentation. *Millennium* 25 (2): 255–289.

Habermas, J. 1984. *The Theory of Communicative Action, Vol. 1.* Boston: Beacon Press.

———. 1987. *The Theory of Communicative Action, Vol. 2.* Boston: Beacon Press.

———. 1992a. Further Reflections on the Public Sphere. In *Habermas and the Public Sphere.* Ed. C. Calhoun, 421–461. Cambridge, MA: MIT Press.

———. 1992b. *Moral Consciousness and Communicative Action.* Cambridge: Polity.

———. 1996. *Between Facts and Norms.* Cambridge: Polity.

———. 1998a. The European Nation-State: On the Past and Future of Sovereignty and Citizenship. *Public Culture* 10 (2): 397–416.

———. 1998b. *The Inclusion of the Other: Essays in Political Theory.* Cambridge: Cambridge University Press.

———. 1999. *The Structural Transformation of the Public Sphere.* Trans. T. Burger and F. Lawrence. Cambridge: Polity.

———. 2000. *The Postnational Constellation.* Cambridge: Polity.

———. 2001. Why Europe Needs a Constitution. *New Left Review* (September/October): 5–26.

Halleck, D. 1994. Zapatistas On-Line. *North American Congress on Latin America: Report on the Americas* 28 (2): 30–33

Hamelink, C. J. 1994. *The Politics of World Communication.* London: Sage.

———. 1998. The People's Communication Charter. *Development in Practice* 8 (1): 68–74.

Harcourt, W., ed. 1999. *Women@Internet.* London: Zed.

———. 2000. *World Wide Women and the Web.* Ed. D. Gauntlett, 150–158. *Web.Studies: Rewiring Media Studies for the Digital Age.* London: Arnold.

Harcourt, W., L. Rabinovich, and F. Alloo. 2002. Women's Networking and Alliance Building: The Politics of Organizing in and Around Place. *Development* 45 (1): 42–47.

Harvey, D. 1989. *The Condition of Postmodernity.* Oxford: Blackwell.

Held, D. 1991. Democracy, the Nation-State and the Global System. In *Political Theory Today.* Ed. D. Held, 197–235. Cambridge: Polity.

———, ed. 1993. *Prospects for Democracy: North, South, East, West.* Cambridge: Polity.

———. 1995. *Democracy and Global Order.* Cambridge: Polity.

———. 2002. Law of States, Law of Peoples. *Legal Theory* 8 (2): 1–44.

———. 2003. Cosmopolitanism: Ideas, Realities and Deficits. In *Governing Globalization.* Ed. D. Held and A. McGrew, 305–324. Cambridge: Polity.

Held, D., and A. McGrew, eds. 2003. *Governing Globalization: Power, Authority and Global Governance.* Cambridge: Polity.

Held, D., A. McGrew, D. Goldblatt, and J. Perraton, eds. 1999. *Global Transformations: Politics, Economics and Culture.* Cambridge: Polity.

Herman, E., and N. Chomsky. 1988. *Manufacturing Consent: The Political Economy of the Mass Media.* New York: Pantheon.

Herman, E., and R. W. McChesney. 1997. *The Global Media: The New Missionaries of Corporate Capitalism.* London: Cassell.

Hill, D. T., and K. Sen. 2000. The Internet in Indonesia's New Democracy. *Democratization* 7 (1): 119–136.

Hill, K. A., and J. E. Hughes. 1998. *Cyberpolitics: Citizen Activism in the Age of the Internet.* Oxford: Rowman and Littlefield.

Hill, M., and W. Montag, eds. 2000. *Masses, Classes and the Public Sphere.* London: Verso.

Hirst, P., and G. Thompson. 1996. *Globalization in Question: The International Economy and the Possibilities of Governance.* Cambridge: Polity.

Hoffman, M. 1987. Critical Theory and the Inter-paradigm Debate. *Millennium* 16 (2): 231–249.

———. 1988. Conversations on International Critical Theory. *Millennium* 17 (1): 91–95.

———. 1991. Restructuring, Reconstruction, Reinscription, Rearticulation: Four Voices in Critical International Theory. *Millennium* 20 (2): 169–185.

———, ed. 1993. *Political Theory, International Relations and the Ethics of Intervention.* London: Macmillan.

Holloway, J., and E. Peláez. 1998. *Zapatista! Reinventing Revolution in Mexico.* London: Pluto Press.

Holton, R. J. 1998. *Globalization and the Nation State.* London: Macmillan.

Holub, R. C. 1991. *Jürgen Habermas: Critic in the Public Sphere.* London: Routledge.

Honneth, A. 1995. *The Struggle for Recognition: The Moral Grammar of Social Conflicts.* Cambridge: Polity.

Hoogvelt, A. 1997. *Globalization and the Postcolonial World: The New Political Economy of Development.* London: Macmillan.

Howard, M., G. J. Andreopoulos, and M. R. Shulman. 1994. *The Laws of War: Constraints on Warfare in the Western World.* New Haven, CT: Yale University Press.

Hsiung, P.-C., and Y.-L. R. Wong. 1998. *Jie Gui*—Connecting the Tracks: Chinese Women's Activism Surrounding the 1995 World Conference on Women in Beijing. *Gender & History* 10 (3): 470–497.

Human Rights Watch, 2007. Technology as a Restraint: Internet Censorship and Surveillance. In *World Report 2007.* Available from http://www.hrw.org/wr2k7/essays/shrinking/5.htm#_Toc152464115 (accessed my 1, 2007).

Hurd, I. 1999. Legitimacy and Authority in International Politics. *International Organization* 53 (2): 379–408.

Hurrell, A. 1995. International Political Theory and the Global Environment. In *International Relations Theory Today.* Ed. K. Booth and S. Smith, 129–153. Cambridge: Polity.

Hurrell, A., and B. Kingsbury. 1992. *The International Politics of the Environment*. Oxford: Clarendon Press.

IFPI. 2006. *The Recording Industry Piracy Report 2006*. International Federation of the Phonographic Industry. Available from http://www.ifpi.org/content/library/piracy-report2006.pdf (accessed May 1, 2007).

ITU. 2003. *World Telecommunication Development Report: Access Indicators for the Information Society*. 6th ed. International Telecommunication Union. Geneva: ITU.

————. 2006a. *ITU Trends in Telecommunications Reform: Measuring ICT for Social and Economic Development*. International Telecommunication Union. 7th ed. Geneva: ITU.

————. 2006b. *World Information Society Report 2006*. International Telecommunication Union. Geneva: ITU.

————. 2006c. *World Telecommunication Development Report 2006*. 7th ed. International Telecommunication Union. Geneva: ITU.

ITU Statistics. Did You Know That...? International Telecommunication Union. Available from http://www.itu.int/ITU-D/ict/statistics/ict/index.html (accessed May 1, 2007).

Jackson, R. 1990. *Quasi-States: Sovereignty, International Relations and the Third World*. Cambridge: Cambridge University Press.

Jackson, R. H., and C. G. Rosberg. 1982. Why Africa's Weak States Persist: The Empirical and the Juridical in Statehood. *World Politics* 35 (1): 1–24.

Jameson, F. 1991. *Postmodernism, Or, the Cultural Logic of Late Capitalism*. London: Verso.

Jensen, M. 2003. ICT in Africa: A Status Report. *Global Information Technology Report: Readiness for the Networked World*. Ed. S. Dutta, B. Lanvin, and F. Paua, 86–101. Oxford: Oxford University Press.

Johnston, J., and G. Laxter. 2003. Solidarity in the Age of Globalisation: Lessons from the Anti-MAI and Zapatista Struggle. *Theory & Society* 32 (1): 39–91.

Jones, S. G., ed. 1995. *CyberSociety: Computer-Mediated Communication and Community*. London: Sage.

Joseph, R. 2001. Understanding the Digital Divide. *Prometheus* 19 (4): 333–336.

Kalathil, S., and T. C. Boas. 2001. *The Internet and State Control in Authoritarian Regimes: China, Cuba and the Counterrevolution*. Washington, DC: Carnegie Endowment for International Peace.

————. 2003. *Open Networks, Closed Regimes: The Impact of the Internet on Authoritarian Regimes*. Washington, DC: Carnegie Endowment for International Peace.

Keane, J. 1984. *Public Life and Late Capitalism: Towards a Socialist Theory of Democracy*. Cambridge: Cambridge University Press.

————. 1998. *Civil Society and the State: New European Perspectives*. Cambridge: Polity.

Kalathil, S., and T. C. Boas. 2001. Global Civil Society? In *Global Civil Society 2001*. Ed. H. Anheier, M. Glasius, and M. Kaldor, 25–47. Oxford: Oxford University Press.

Keck, M. E., and K. Sikkink. 1998. *Activists Beyond Borders: Advocacy Networks in International Politics*. Ithaca, NY: Cornell University Press.

Kennedy, T. L. M. 2000. An Exploratory Study of Feminist Experiences in Cyberspaces. *Cyberpsychology & Behaviour* 3 (5): 707–719.

Keohane, R. O. 1995. Hobbes' Dilemma and Institutional Change in World Politics: Sovereignty in International Society. In *Whose World Order?* Ed. H. H. Holm and G. Sorenson, 165–186. Boulder, CO: Westview Press.

———. 1998. Power and Interdependence in the Information Age. *Foreign Affairs* 77 (5): 81–94.

Keohane, R. O., and S. Hoffman, eds. 1990. *The New European Community*. Oxford: Westview Press.

Keohane, R. O., and J. S. Nye. 1972. *Transnational Relations and World Politics*. New York: Longman.

———. 1977. *Power and Interdependence: World Politics in Transition*. Boston: Little & Brown.

———. 1998. Power and Interdependence in the Information Age. *Foreign Affairs* 77 (5): 81–94.

Knouse, S. B., and S. C. Webb. 2001. Virtual Networking for Women and Majorities. *Career Development International* 6 (4): 226–229.

Knudson, T. J. 1992. *A History of International Relations Theory*. Manchester, UK: Manchester University Press.

Koenig-Archibugi, M. 2003. Mapping Global Governance. In *Governing Globalization: Power, Authority and Global Governance*. Ed. D. Held and A. McGrew, 46–69. Cambridge: Polity.

Köhler, M. 1998. From the National to the Cosmopolitan Public Sphere. In *Re-imagining Political Community*. Ed. D. Archibugi, D. Held, and M. Köhler, 231–251. Cambridge: Polity.

Koopmans, R., and J. Erbe. 2004. Towards a European Public Sphere? Vertical and Horizontal Dimensions of Europeanized Political Communication. *Innovation* 17 (2): 97–118.

Koselleck, R. 1988. *Critique and Crisis: Enlightenment and the Pathogenesis of Modern Society*. Cambridge, MA: MIT Press.

Krasner, S. D. 1982. Structural Consequences and Regime Consequences. *International Organization* 36 (2): 185–206.

———. 1991. Global Communications and National Power: Life on the Pareto Frontier. *World Politics* 43 (3): 336–366.

Kratochwil, F. 1989. *Norms, Rules and Decisions*. Cambridge: Cambridge University Press.

Landes, J. B., ed. 1988. *Women and the Public Sphere in the Age of the French Revolution*. Oxford: Oxford University Press.

———. 1998. *Feminism, the Public and the Private*. Oxford: Oxford University Press.

Lash, S., and M. Featherstone. 2001. Recognition and Difference. *Theory, Culture and Society* 18 (2–3): 1–20.

Lessig, L. 1999. *Code and Other Laws of Cyberspace*. New York: Basic Books.

Liberman, P. 1993. The Spoils of Conquest. *International Security* 18 (2): 125–153.

Linebaugh, P., and M. Rediker. 2000. *The Many-Headed Hydra: Sailors, Slaves, Commoners, and the Hidden History of the Revolutionary Atlantic*. Boston: Beacon Press.

Linklater, A. 1990a. *Beyond Realism and Marxism: Critical Theory and International Relations*. London: Macmillan.

———. 1990b. *Men and Citizens in the Theory of International Relations*. London: Macmillan.

———. 1990c. The Problem of Community in International Relations. *Alternatives* 15 (2): 135–153.

———. 1992. The Question of the Next Stage in International Relations Theory: A Critical-Theoretical Point of View. *Millennium* 21 (1): 77–98.

———. 1994. Dialogue, Dialectic and Emancipation in International Relations at the End of the Post-War Age. *Millennium* 23 (2): 119–131.

———. 1996a. The Achievements of Critical Theory. In *International Theory: Positivism and Beyond*. Ed. S. Smith, K. Booth, and M. Zalewski, 279–298. Cambridge: Cambridge University Press.

———. 1996b. Citizenship and Sovereignty in the Post-Westphalian State. *European Journal of International Relations* 2 (1): 77–103.

———. 1998. *The Transformation of Political Community*. Columbia: University of South Carolina Press.

Lipschutz, R. 1992. Reconstructing World Politics: The Emergence of Global Civil Society. *Millennium* 21 (3): 398–420.

Lucas, R. E. 2000. Some Macroeconomics for the 21st Century. *Journal of Economic Perspectives* 14 (1): 159–168.

Lucas Jr., H. C., and R. Sylla. 2003. The Global Impact of the Internet: Widening the Economic Gap between Wealthy and Poor Nations. *Prometheus* 21 (1): 3–21.

Lynch, M. 1999. *State Interests and Public Spheres: The International Politics of Jordan's Identity*. New York: Columbia University Press.

———. 2000. The Dialogue of Civilizations and International Public Spheres. *Millennium* 29 (2): 307–330.

———. 2002. Why Engage? China and the Logic of Communicative Engagement. *European Journal of International Relations* 8 (2): 187–230.

———. 2003. Taking Arabs Seriously. *Foreign Affairs* 82 (5): 81–94.

———. 2005. Transnational Dialogue in an Age of Terror. *Global Society* 19 (1): 5–28.

Lyotard, J.-F. 1984. *The Postmodern Condition: A Report on Knowledge*. Manchester, UK: Manchester University Press.

Mahieu, L. 1988. *A Culture for Democracy*. Oxford: Clarendon Press.

Main, L. 2001. The Global Information Infrastructure: Empowerment or Imperialism? *Third World Quarterly* 22 (1): 83–97.

Mancini, G. 1990. The Making of a Constitution for Europe. In *The New European Community*. Ed. R. O. Keohane and S. Hoffman, 177–194. Oxford: Westview Press.

Mancusi-Materi, E. 1999. Environmental Concerns and Virtual Politics: Knowledge and Activism on the Web. *Development* 42 (2): 74–82.

Mbambo, B. 1999. Disseminating African Women's Information on the Internet: Issues and Constraints. *Information Development* 15 (2): 103.

McAdam, D., S. Tarrow, and C. Tilly. 2000. *Dynamics of Contention*. Cambridge: Cambridge University Press.

McChesney, R. W. 2001. Global Media, Neoliberalism and Imperialism. *Monthly Review* 52 (1). Available from http://www.monthlyreview.org/301rwm.htm (accessed May 1, 2007).

McGrew, A. 1998. The Globalisation Debate: Putting the Advanced Capitalist State in its Place. *Global Society* 12 (3): 299–319.

McLaughlin, L. 1993. Feminism, the Public Sphere, Media and Democracy. *Media, Culture & Society* 15 (3): 599–620.

Meehan, J., ed. 1995. *Feminists Read Habermas*. London: Routledge.

Meldrum, A. 2004. Mugabe Introduces New Curbs on the Internet. *The Guardian*. May 3: 17.

Melucci, A. 1996. *Challenging Codes: Collective Action in the Information Age*. Cambridge: Cambridge University Press.

Meyrowitz, J. 1985. *No Sense of Place*. Oxford: Oxford University Press.

Miller, M. C. 2002. What's Wrong with This Picture? *The Nation*. January 7: 13.

Miller, S. 1996. *Civilizing Cyberspace: Policy, Power and the Information Superhighway*. New York: ACM Press.

Mitra, A. 2001. Marginal Voices in Cyberspace. *New Media & Society* 3 (1): 29–48.

Mitzen, J. 2001. Towards a Visible Hand: The International Public Sphere in Theory and Practice. Diss., Chicago: University of Chicago.

———. 2005. Reading Habermas in Anarchy: Multilateral Diplomacy and Global Public Spheres. *American Political Science Review* 99 (3): 401–417.

Moghadan, V. M. 2005. *Globalizing Women: Transnational Feminist Networks*. Baltimore, MD: Johns Hopkins University Press.

Mohammadi, A., ed. 1997. *International Communication and Globalization*. London: Sage.

Molina, A. 2003. The Digital Divide: The Need for a Global e-Inclusion Movement. *Technology Analysis and Strategic Management* 15 (1): 138–152.

Morgenthau, H. J. 1948. *Politics among Nations*. New York: Alfred A. Knopf.

Morton, A. D. 2002. La Resurrecion del Mais: Globalisation, Resistance and the Zapatistas. *Millennium* 31 (1): 27–54.

MPAA. 2006. The Cost of Movie Piracy. Motion Picture Association of America. Available from http://www.mpaa.org/2006_05_03leksumm.pdf (accessed May 1, 2007).

Murdoch, R. 2005. Speech by Rupert Murdoch to the American Society of Newspaper Editors. News Corporation. Press Release, April 13. Available from http://www.newscorp.com/news/news_247.html (accessed May 1, 2007).

Murphy, A. B. 1996. The Sovereign State System as a Political-Territorial Ideal: Historical and Contemporary Considerations. In *State Sovereignty as a Social Construct*. Ed. T. J. Biersteker and C. Weber, 81–121. Cambridge: Cambridge University Press.

Murphy, C. 2000. Global Governance: Poorly Done and Poorly Understood. *International Affairs* 76 (4): 789–803.

Naughton, J. 1999. *A Brief History of the Future*. London: Weidenfeld and Nicolson.

Negroponte, N. 1995. *Being Digital*. New York: Knopf.

Negt, O., and A. Kluge. 1993. *Public Sphere and Experience: Toward an Analysis of the Bourgeois and Proletarian Public Sphere*. Minneapolis: University of Minneapolis Press.

Netcraft. 2007. May 2007 Web Server Survey. Available from http://news.netcraft.com/archives/2007/05/01/may_2007_web_server_survey.html (accessed May 11, 2007)

Neufeld, M. 1995. *The Restructuring of International Relations Theory*. Cambridge: Cambridge University Press.

New Internationalist. 2001. Tomorrow Begins Today: The Zapatista Declaration That Launched a Global Movement. *New Internationalist* 338 (September): 14–15.

Nickel, J. W. 2002. Is Today's International Human Rights System a Global Governance Regime? *The Journal of Ethics* 6 (4): 353–371.

Nip, J. Y. M. 2004. The Relationship between Online and Offline Communities: The Case of the Queer Sisters. *Media, Culture & Society* 26 (3): 409–428.

Norris, P. 2001. *Digital Divide: Civic Engagement, Information Poverty, and the Internet Worldwide*. Cambridge: Cambridge University Press.

Norton, R. 1992. *Mother Clap's Molly House: The Gay Subculture in England, 1700–1830*. London: Gay Men's Press.

Noveck, B. S. 2000. Paradoxical Partners: Electronic Communication and Electronic Democracy. *Democratization* 7 (1): 18–35.

NTIA. 1999. *Falling through the Net: Defining the Digital Divide*. National Telecommunications and Information Administration. Washington: U.S. Department of Commerce.

O'Donnell, S. 2001. Analysing the Internet and the Public Sphere: The Case of *Womenslink*. *Javnost - The Public* 8 (1): 39–58.

OECD. 2000. Understanding the Digital Divide. Organization for Economic Cooperation and Development. Paris: OECD.

———. 2006. Are Students Ready for a Technology-Rich World? Organization for Economic Cooperation and Development. Available from http://www.oecd.org/dataoecd/28/4/35995145.pdf (accessed May 1, 2007).

Ohmae, K. 1995. *The End of the Nation State*. New York: Free Press.

Olesen, T. 2004. Globalising the Zapatistas: From Third World Solidarity to Global Solidarity. *Third World Quarterly* 25 (1): 255–267.

OneStat. 2007. Microsoft's Internet Explorer Global Usage Share Is 85.81 Percent According to OneStat.com. Available from http://www.onestat.com/html/aboutus_pressbox50-microsoft-internet-explorer-7-usage.html (accessed May 1, 2007).

OpenNet Initiative. 2004. Internet Filtering in Saudi Arabia in 2004. Available from http://www.opennetinitiative.net/studies/saudi/ (accessed May 1, 2007).

———. 2005a. Internet Filtering in China in 2004–2005: A Country Study. Available from http://www.opennetinitiative.net/studies/china/ (accessed May 1, 2007).

———. 2005b. Internet Filtering in Iran in 2004–2005: A Country Study. Available from http://www.opennetinitiative.net/studies/iran/ (accessed May 1, 2007).

Ott, D., and M. Rosser. 2000. The Electronic Republic? The Role of the Internet in Promoting Democracy in Africa. *Democratization* 7 (1): 137–155.

Parmentier, R. 1999. Greenpeace and the Dumping of Waste at Sea: A Case of Non-State Actors. *International Negotiation* 4 (3): 435–457.

Pasha, M. K., and D. L. Blaney. 2001. Elusive Paradise: The Promise and Peril of Global Civil Society. *Alternatives* 23 (3): 417–430.

Pateman, C. 1987. Feminist Critiques of the Public/Private Dichotomy in Disorder of Women. In *Feminist Critiques*. Ed. A. Phillips, 103–126. Oxford: Blackwell.

———. 1988. The Fraternal Social Contract. In *Civil Society and the State*. Ed. J. Keane, 101–127. London: Verso.

Paterson, M. 2001. *Understanding Environmental Politics: Domination, Accumulation, Resistance*. Basingstoke, UK: Macmillan.

Peters, J. D. 1993. Distrust of Representation: Habermas on the Public Sphere. *Media, Culture & Society* 15 (4): 541–571.

Peterson, V. S. 1992. *Gendered States: Feminist (Re)visions of International Relations Theory*. Boulder, CO: Lynne Rienner.

Pew Internet. Demographics of Internet Users. Pew Internet and American Life Project. Available from http://www.pewinternet.org/trends/User_Demo_4.26.07.htm (accessed May 11, 2007).

Phillips, A. 1998. Dealing with Difference: A Politics of Ideas or a Politics of Presence? In *Feminism: The Public and the Private*. Ed. J. B. Landes, 475–495. Oxford: Oxford University Press.

Philpott, D. 1999. Westphalia, Authority, and International Society. *Political Studies* 47 (3): 566–589.

Pickerill, J. 2003. *Cyberprotest: Environmental Activism Online*. Manchester, UK: Manchester University Press.

Pierre, J., and B. G. Peters. 2000. *Governance, Politics and the State*. London: Palgrave.

Poster, M. 1989. *Critical Theory and Poststructuralism: In Search of a Context.* Ithaca, NY: Cornell University Press.

———. 1995a. CyberDemocracy: Internet and the Public Sphere. Irvine: University of California. Available from http://www.hnet.uci.edu/ mposter/writings/democ.html (accessed May 1, 2007).

———. 1995b. The Net as a Public Sphere? Wired. Available from http:// www.wired.com/wired/archive/3.11/poster.if_pr.html (accessed May 1, 2007).

Price, R. 1998. Anarchy in International Relations Theory: The Neorealist-Neoliberal Debate. *International Organization* 48 (2): 313–334.

Reinicke, W. H. 1999. The Other World Wide Web: Global Public Policy Networks. *Foreign Policy* 117(Winter): 44–57.

Rendall, S., and Broughel, T. 2003. Amplifying Officials, Squelching Dissent. *Extra!* May/June. Fairness and Accuracy in Reporting. Available from http://www.fair.org/index.php?page=1145 (accessed May 1, 2007).

Rengger, N. 1988. Going Critical? A Response to Hoffman. *Millennium* 17 (1): 81–89.

Reporters Without Borders. 2004. China. In Internet under Surveillance. Available from http://www.rsf.org/article.php3?id_article=10749 (accessed May 1, 2007).

———. 2005. *Handbook for Bloggers and Cyber-Dissidents.* Available from http://www.rsf.org/IMG/pdf/handbook_bloggers_cyberdissidents-GB. pdf (accessed May 1, 2007).

———. 2007. Internet in 2007. In *Annual Report 2007.* Available from http:// www.rsf.org/rubrique.php3?id_rubrique=675 (accessed May 1, 2007).

RIAA. 2006. RIAA Identifies 12 Piracy "Hot Spot" Cities. Recording Industry Association of America. Press Release, May 3. Available from http://www. riaa.com/News/newsletter/050306.asp (accessed May 1, 2007).

Richards, A., and M. Schnall. 2003. Cyberfeminism: Networking on the Net. Extract from *Sisterhood Is Forever: A Women's Anthology for a New Millennium.* Ed. R. Morgan. New York: Washington Square Press. Available from http://www.feminist.com/resources/artspeech/genwom/ cyberfeminism.html (accessed May 1, 2007).

Risse, T. 2000. "Let's Argue!" Communicative Action in World Politics. *International Organization* 54 (1): 1–39.

Robertson, R. 1992. *Globalization: Social Theory and Global Culture.* London: Sage.

Rosenau, J. N. 1990. *Turbulence in World Politics: A Theory of Change and Continuity.* Princeton: Princeton University Press.

———. 1997. *Along the Domestic-Frontier. Exploring Governance in a Turbulent World.* Cambridge: Cambridge University Press.

———. 2003. Governance in a New Global Order. In *Governing Globalization: Power, Authority and Global Governance.* Ed. D. Held and A. McGrew, 70–86. Cambridge: Polity.

Rosenau, J. N. 2005. Global Governance as Disaggregated Complexity. In *Contending Perspectives on Global Governance*. Ed. A. D. Ba and M. J. Hoffman, 131–153. London: Routledge.

Rosenau, J. N., and E.-O. Czempiel, eds. 1992. *Governance Without Government: Order and Change in World Politics*. Cambridge: Cambridge University Press.

Rosenburg, J. 1990. What's the Matter with Realism? *Review of International Studies* 16 (4): 296–299.

Routledge, P. 1998. Going Globile: Spatiality, Embodiment and Mediation in the Zapatista Insurgency. In *Rethinking Geopolitics*. Ed. S. Dalby and G. O' Tuathall, 240–260. London: Routledge.

———. 2001. "Our Resistance Will Be as Transnational as Capital": Convergence Space and Strategy in Globalising Resistance. *GeoJournal* 52 (1): 25–33.

Ruggie, J. G. 1993. Territoriality and Beyond: Problematizing Modernity in International Relations. *International Organization* 47 (1): 139–174.

Russell, A. 2001a. Chiapas and the NEW News: Internet and Newspaper Coverage of a Broken Cease-Fire. *Journalism* 2 (2): 7–29.

———. 2001b. The Zapatistas Online: Shifting the Discourse of Globalization. *Gazette* 63 (5): 399–413.

Russett, B. 1993. *Grasping the Democratic Peace: Principles for a Post–Cold War World*. Princeton: Princeton University Press.

Ryan, M. P. 1992. Gender and Public Access: Women's Politics in Nineteenth Century America. In *Habermas and the Public Sphere*. Ed. C. Calhoun, 259–288. Cambridge, MA: MIT Press.

Salter, L. 2004. Structure and Forms of Use: A Contribution to Understanding the "Effects" on the Internet of Deliberative Democracy. *Information, Communication and Society* 7 (2): 185–206.

Sassi, S. 2001. The Transformation of the Public Sphere? In *New Media and Politics*. Ed. B. Axford and R. Huggins, 89–108. London: Sage.

Saurin, J. 1995. The End of International Relations? The State and International Relations Theory in the Age of Globalisation. In *Boundaries in Question: New Directions in International Relations*. Ed. J. Macmillan and A. Linklater, 244–261. London: Pinter.

Scannell, P. 1989. Public Service Broadcasting and Modern Public Life. *Media, Culture & Society* 11 (2): 135–166.

Scherer, J. 1972. *Contemporary Community: Sociological Illusion or Reality?* London: Tavistock.

Schermers, H. G., and N. M. Blokker. 1995. *International Institutional Law*. 3rd ed. The Hague: Martinus Nijhoff.

Schnall, M. A Decade on the Internet Serving Women and Girls. Feminist. com. Available from http://www.feminist.com/about/whatis.html (accessed May 1, 2007).

Scholte, J. A. 2000. *Globalization: A Critical Introduction*. London: Macmillan.

Schudson, M. 1992. Was There Ever a Public Sphere? In *Habermas and the Public Sphere*. Ed. C. Calhoun, 143–163. Cambridge, MA: MIT Press.

Shade, L. R. 1996. Is There Free Speech on the Net? Censorship in the Global Information Infrastructure. In *Cultures of Internet*. Ed. R. Shields, 11–32. London: Sage.

Shields, R., ed. 1996. *Cultures of Internet: Virtual Spaces, Real Histories, Living Bodies*. London: Sage.

Simpson, S. 2004. Explaining the Commercialization of the Internet. A Neo-Gramscian Contribution. *Information, Communication & Society 7* (1): 50–68.

Slaughter, A.-M. 1997. The Real New World. *Foreign Affairs* 76 (5): 183–198.

Slevin, J. 2000. *The Internet and Society*. Cambridge: Polity.

Smith, A. D. 1990. Towards a Global Culture? In *Global Culture: Nationalism, Globalization and Modernity*. Ed. M. Featherstone, 171–191. London: Sage.

———. 1995. *Nations and Nationalism in a Global Era*. Cambridge: Polity.

Smith, D. 2006. Anywhere, Anytime: Wi-Fi's the Future. In Special Report Section: The Broadband Revolution. *The Observer*, December 10: 2.

Smith, E. 2000. *Mexican Women's Movement Makes Internet Work for Women*. Connected: Women & the Internet. Available from http://www. connected.org/women/erika.html (accessed May 1, 2007).

Smith, S. 2000. The Increasing Insecurity of Security Studies: Conceptualizing Security in the Last Twenty Years. In *Critical Reflections on Security and Change*. Ed. S. Croft and T. Terriff, 72–106. London: Frank Cass & Co.

Smith, M. A., and P. Kollock, eds. 1999. *Communities in Cyberspace*. London: Routledge.

Sobel, D., ed. 1999. *Filters and Freedom: Free Speech Perspectives on Internet Content Controls*. Washington, DC: Electronic Privacy Information Center.

Sparks, C. 1998. A Global Public Sphere? In *Electronic Empires*. Ed. D. K. Thussu, 108–124. London: Arnold.

———. 2001. The Internet and the Global Public Sphere. In *Mediated Politics*. Ed. L. W. Bennett and R. Entman, 75–95. Cambridge: Cambridge University Press.

Stahler-Sholk, R. 2001. Globalization and Social Movement Resistance: The Zapatista Rebellion in Chiapas, Mexico. *New Political Science* 23 (4): 493–516.

Starr, A. 2000. *Naming the Enemy: Anti-corporate Movements Confront Globalization*. London: Zed Books.

Steans, J. 2003. Global Governance: A Feminist Perspective. In *Governing Globalization: Power, Authority and Global Governance*. Ed. D. Held and A. McGrew, 87–110. Cambridge: Polity.

Stevenson, N. 1993. *Understanding Media Cultures*. London: Sage.

Stevenson, R. 1992. Defining International Communication as a Field. *Journalism Quarterly* 69 (3): 543–553.

Stewart, A. 2001. *Theories of Power and Domination: The Politics of Empowerment in Late Modernity.* London: Sage.

Stone, A. R. 1991. Will the Real Body Please Stand Up? Boundary Stories about Virtual Cultures. In *Cyberspace.* Ed. M. Benedikt, 81–118. Cambridge, MA: MIT Press.

Strange, S. 1996. *The Retreat of the State: The Diffusion of Power in the World Economy.* Cambridge: Cambridge University Press.

Street, J. 2001. *Mass Media, Politics and Democracy.* New York: Palgrave.

Sunstein, C. 2001. *Republic.com.* Princeton: Princeton University Press.

Suter, K. 2005. The International Court of Justice at Sixty. *Contemporary Review* 287 (1676): 124–134.

Swett, C. 1995. Strategic Assessment: The Internet. Project on Government Secrecy. Federation of American Scientists. Available from http://www.fas.org/cp/swett.html (accessed May 1, 2007).

Sylvester, C. 1994. *Feminist Theory and International Relations in a Postmodern Era.* Cambridge: Cambridge University Press.

Tait, R. 2006. Censorship Fears Rise as Iran Blocks Access to Top Websites. *The Guardian,* December 4. Available from http://www.guardian.co.uk/iran/story/0,,1963166,00.html (accessed May 1, 2007).

Tanzi, V. 1995. *Taxation in an Integrating World.* Washington, DC: Brookings Institution.

Tarrow, S. 1998. *Power in Movement: Social Movements and Contentious Politics.* 2nd ed. Cambridge: Cambridge University Press.

Taylor, C. 1994. The Politics of Recognition. In *Multiculturalism: Examining the Politics of Recognition.* Ed. A. Gutmann, 25–74. Princeton: Princeton University Press.

Tehranian, M. 1999. *Global Communication and World Politics: Domination, Development, and Discourse.* London: Lynne Rienner.

Thierer, A. 2002. *190* Internet Censors? Rising Threats to Online Speech. CATO Institute. Available from http://www.cato.org/tech/tk/020726-tk.html (accessed May 1, 2007).

Thomas, C. 1997. New Myths for the South: Globalization and the Conflict between Private Power and Freedom. In *The South in Global Politics.* Ed. C. Thomas and P. Wilkin, 1–17. London: Macmillan.

———. 2000. *Global Governance, Development and Human Security.* London: Pluto Press.

Thompson, J. B. 1990. *Ideology and Modern Culture: Critical Social Theory in the Era of Mass Communication.* Cambridge: Polity.

———. 1995. *The Media and Modernity: A Social Theory of the Media.* Cambridge: Polity Press.

Thussu, D. K. 2006. *International Communication: Continuity and Change.* 2nd ed. London: Arnold.

Tickner, J. A. 1992a. *Gender in International Relations: Feminist Perspectives on Achieving Global Security.* New York: Columbia University Press.

———. 1992b. You Just Don't Understand. Troubled Engagements between Feminists and International Relations Theorists. *International Organization* 41 (4): 611–652.

Tomlinson, J. 1999. *Globalization and Culture.* Cambridge: Polity.

Tonnies, F. 1957. *Community and Society.* Lansing: Michigan State University Press.

Tsoukas, H. 1999. David and Goliath in the Risk Society: Making Sense of the Conflict between Shell and Greenpeace in the North Sea. *Organization* 6 (3): 499–528.

UIA, eds. 2007. *Yearbook of International Associations 2006/7.* Brussels: Union of International Associations.

UN. 1988 *Human Rights: A Compilation of International Instruments.* New York: United Nations.

———. 1992. United Nations Framework Convention on Climate Change. United Nations. Available from http://unfccc.int/essential_background/convention/background/items/2853.php (accessed May 1, 2007).

———. 2000. United Nations Millennium Declaration. United Nations. Available from http://www.un.org/millennium/declaration/ares552e.htm (accessed May 1, 2007).

UNCTAD, 2006a. *The Digital Divide Report: ICT Diffusion Index 2005.* United Nations Conference on Trade and Development. Geneva: United Nations.

———. 2006b. *Information Economy Report 2006.* United Nations Conference on Trade and Development. Geneva: United Nations.

UNDP. 2001. *Human Development Report 2001: Making New Technologies Work for Human Development.* United Nations Development Programme. Available from http://hdr.undp.org/reports/global/2001/en/ (accessed May 1, 2007).

———. 2004. *Promoting ICT for Human Development in Asia.* United Nations Development Programme. Available from http://hdr.undp.org/docs/reports/regional/ASP_ASIA_PACIFIC/South_East_Asia_2005_en.pdf (accessed May 1, 2007).

UNESCO Institute of Statistics, eds. 2005. Measuring Linguistic Diversity on the Internet. United Nations Educational, Scientific and Cultural Organization. Montreal, Canada: UNESCO.

Valdivia, A. N. 1995. *Feminism, Multiculturalism and the Media: Global Diversities.* London: Sage.

van Zoonen, L. 1991. A Tyranny of Intimacy? Women, Femininity and Television News. In *Communication and Citizenship: Journalism and the Public Sphere in the New Media Age.* Ed. P. Dahlgren and C. Sparks, 217–235. London: Routledge.

Väyrynen, R., ed. 1999. *Globalization and Global Governance.* New York: Rowman and Littlefield.

Verstraeten, H. 1996. The Media and the Transformation of the Public Sphere: A Contribution for a Critical Political Economy of the Public Sphere. *European Journal of Communication* 11 (3): 347–371.

Vincent, J. 1992. Modernity and Universal Human Rights. In *Global Politics: Globalization and the Nation State*. Ed. A. McGrew and P. G. Lewis, 272–280. Cambridge: Cambridge University Press.

Virnoche, M. E., and G. T. Marx. 1997. "Only Connect": E. M. Forster in an Age of Electronic Communication: Computer-Mediated Association and Community Networks. *Sociological Inquiry* 67 (1): 23–37.

Walker, R. B. J. 1991. State Sovereignty and the Articulation of Political Space/Time. *Millennium* 20 (3): 445–462.

———. 1993. *Inside/Outside: International Relations as a Political Theory*. Cambridge: Cambridge University Press.

Wallace, W. 1994. Rescue or Retreat? The Nation State in Western Europe. *Political Studies* 42: 52–76.

Wallerstein, I. 1977. *The Capitalist World Economy: Essays by Immanuel Wallerstein*. Cambridge: Cambridge University Press.

Waltz, K. W. 1979. *Theory of International Politics*. Reading: Addison-Wesley.

Wapner, P. 2002. Horizontal Politics: Transnational Environmental Activism and Global Cultural Change. *Global Environmental Politics* 2 (2): 37–62.

Ward, K. J. 1999. The Cyber-Ethnographic (Re)Construction of Two Feminist Online Communities. *Sociological Research Online* 4 (1): 1–16.

Warf, B., and J. Grimes, J. 1997. Counterhegemonic Discourses and the Internet. *Geographical Review* 87 (2): 259–274.

Warkentin, C. 2001. *Reshaping World Politics: NGOs, the Internet and Global Civil Society*. Boulder, CO: Rowman and Littlefield.

Warner, M. 1992. The Mass Public and the Mass Subject. In *Habermas and the Public Sphere*. Ed. C. Calhoun, 377–401. Cambridge, MA: MIT Press.

———. 2002. *Publics and Counterpublics*. New York: Zone Books.

Warschauer, M. 2002. Reconceptualizing the Digital Divide. *First Monday* 7 (7). Available from http://firstmonday.org/issues/issue7_7/warschauer/index.html (accessed May 1, 2007).

———. 2003. *Technology and Social Inclusion: Rethinking the Digital Divide*. Cambridge, MA: MIT Press.

Watson, N. 1997. Why We Argue about Virtual Community: A Case Study of the Phish.Net Fan Community. In *Virtual Culture: Identity and Community in Cyberspace*. Ed. S. G. Jones, 102–132. London: Sage.

Watts, J. 2006. Backlash as Google Shores Up Great Firewall of China. *The Guardian*, January 25: 3.

Weber, C. 1992. Reconsidering Statehood: Examining the Sovereignty/Intervention Boundary. *Review of International Studies* 18 (3): 199–216.

Weinburg, J. 1997. Rating the Net. *Hastings Communications and Law Journal* 19 (2): 453–482.

Wendt, A. 1999. *Social Theory of International Politics*. Cambridge: Cambridge University Press.

———. 2001. What Is International Relations For? Notes Towards a Postcritical View. In *Critical Theory and World Politics.* Ed. R. Wyn Jones, 205–224. London: Lynne Rienner.

Wheeler, N. 2002. *Saving Strangers: Humanitarian Intervention in International Society.* Oxford: Oxford University Press

White, C. 2000. Environmental Activism and the Internet. Diss., Albany, New Zealand: Massey University. Available from http://arachna.co.nz/thesis/Chapters.asp?Chapt=5&PageNo=2 (accessed May 1, 2007).

Wight, M. 1966. Why There Is No International Theory? In *Diplomatic Investigations.* Ed. H. Butterfield and M. Wight, 17–34. Cambridge: Cambridge University Press.

Wilkin, P. 2001. *The Political Economy of Global Communication.* London: Pluto Press.

Wired. 1998. A Rebel Movement's Life on the Web. *Wired News.* March 6. Available from http://www.wired.com/politics/law/news/1998/03/10769 (accessed May 1, 2007).

Women-Media Archive. United Nations. Available from http://sdnhq.undp.org/ww/women-media (accessed June 24, 2003).

Women-Media List Archive, a. Networks and Networking. United Nations. Available from http://www.sdnp.undp.org/ww/women-media/msg00282.html (accessed June 24, 2003).

Women-Media List Archive, b. Re: Networks and Networking. United Nations. Available from http://www.sdnp.org/ww/women-media/msg00283.html (accessed June 24, 2003).

Women-Rights List Archive, c. Re: Women's Rights in Globalization. United Nations. Available from http://www.sdnp.undp.org/ww/women-rights/msg00089.html (accessed June 24, 2003).

Womenwatch. Women, the Information Revolution and the Beijing Conference: How Are Women Using the Information Superhighway? United Nations. Available from http://www.un.org/womenwatch/daw/public/w2part2.htm (accessed May 1, 2007).

Woods, N. 2001. Making the IMF and the World Bank More Accountable. *International Affairs* 77 (1): 83–100.

———. 2003. Global Governance and the Role of Institutions. In *Governing Globalization: Power, Authority and Global Governance.* Ed. D. Held and A. McGrew, 25–45. Cambridge: Polity.

World Bank. 2006. *Information and Communication for Development 2006: Global Trends and Policies.* Washington, DC: World Bank.

World Summit on the Information Society. 2005. Tunis Commitment. International Telecommunications Union. Available from http://www.itu.int/wsis/docs2/tunis/off/7.html (accessed May 1, 2007).

Wyn Jones, R., ed. 2001. *Critical Theory and World Politics.* London: Lynne Rienner.

Young, I. M. 1990. *Justice and the Politics of Difference.* Princeton: Princeton University Press.

Young, I. M. 1996. Communication and the Other: Beyond Deliberative Democracy. In *Democracy and Difference*. Ed. S. Benhabib, 120–135. Princeton: Princeton University Press.

Young, O. 1989. *International Regimes*. Ithaca, NY: Cornell University Press.

Youngs, G. 2000. Women Breaking Boundaries in Cyberspace. *Asian Women* 10: 1–18.

———. 2002. Closing the Gaps: Women, Communications and Technology. *Development* 45 (4): 23–28.

Zacher, M. W., and R. A. Matthew. 1995. Liberal International Theory: Common Threads, Divergent Strands. In *Controversies in International Relations Theory*. Ed. C. W. Kegley Jr, 107–150. Basingstoke, UK: Macmillan.

Zalewski, M. 1994. The Women/"Women" Question in International Relations. *Millennium* 23 (2): 407–423.

Zittrain, J., and B. Edelman. 2003a. *Documentation of Internet Filtering in Saudi Arabia*. Cambridge, MA: Berkman Center for Internet & Society, Harvard Law School. Available from http://cyber.law.harvard.edu/filtering/saudiarabia/ (accessed May 1, 2007).

———. 2003b. *Empirical Analysis of Internet Filtering in China*. Cambridge, MA: Berkman Center for Internet & Society, Harvard Law School. Available from http://cyber.law.harvard.edu/filtering/china/ (accessed May 1, 2007).

Zürn, M. 2005. Global Governance and Legitimacy Problems. In *Global Governance and Public Accountability*. Ed. D. Held and M. Koenig-Archibugi, 136–163. Oxford: Blackwell.

INDEX